Biotech's

DICTIONARY *Of* Plant Breeding and Genetics

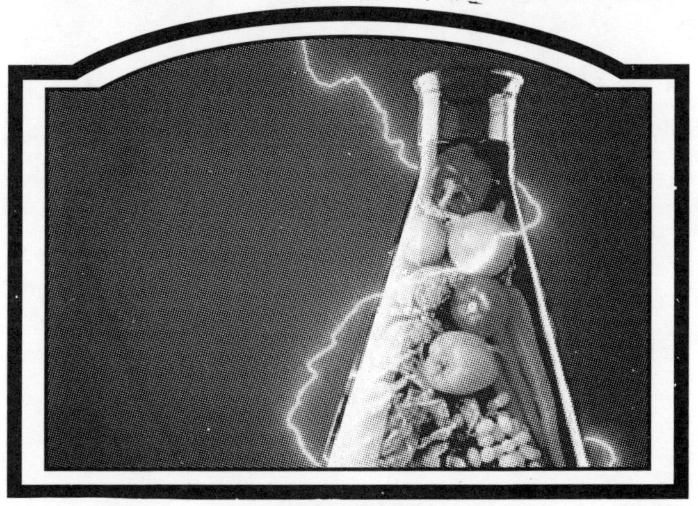

Compiled & Edited by
J.B Jain
&
Sumit Jain

2013
BIOTECH BOOKS
DELHI - 110035

Biotech's

DICTIONARY

Of

Plant Breeding and Genetics

Compiled & Edited by
J.B Jain & Sumit Jain

ISBN 81-7622-141-4

Published by : BIOTECH BOOKS
1123/74, Tri Nagar,
Delhi - 110 035
Phone: 27382765
e-mail: biotechbooks@yahoo.co.in

Showroom : 4762-63/23, Ansari Road, Darya Ganj,
New Delhi - 110 002
Phone: 23245578, 23244987

Printed at : Chawla Offset Printers
New Delhi - 110 052

PRINTED IN INDIA

PREFACE

The cultivation of plants is an important aspect of today's society in many ways. Not only do plants supply us with a major food resource and flow of nutrition but they are also an important source of chemicals and other non-food products such as drugs, oils, latex, pigments and resins. Because of the high value of plants, farmers are constantly trying to increase the yield and quality of their products by using more effective production techniques.

The physical appearance of plants is determined by genetic factors that are inherited from parental crops. The natural transfer of genes can be altered using many genetic techniques such as selection or genetic engineering. One of the important goals of plant breeding is maintaining genetic diversity. This can be achieved through selection and hybridisation. More recently, new techniques have evolved to allow us to grow crops artificially in culture from plant parts to produce whole plants.

Modern science has allowed us to get away from the traditional method of plant breeding to more artificial asexual methods. This allows plants to be produced using individual, or groups of cells, which grow into full plants. Genetics has also allowed us to modify our crops in a more scientific way to select for traits that are more desired by society. In this dictionary, we will be discussing the terms that are related to the Plant Breeding and Genetics.

- **a line**
 the seed-bearing parent line used to produce hybrid seed that is male sterile.
- **abiotic**
 factors or processes of the non-living environment (climate, geology, atmosphere.)
- **abjection**
 the separating of a spore from a sporophore or sterigma by a fungus.
- **abjunction**
 the cutting off of a spore from a hypha by a septum.
- **ablastous**
 without germ or bud.
- **abnormal**
 unusual variance from the natural habit.
- **abort**
 to fail in the early stages of formation, the collapse or disappearance of seeds or cells.
- **abscise**
 separate by abscission, as a leaf from a stem.
- **abscission**
 rejection of plant organs (e.g., of leaves in autumn).
- **absorption**
 uptake of substances, usually nutrients, water or light, by plant cells or tissue, in soil science, the physical uptake of water or ions by a (soil) substance , in microscopy, the interaction of light with matter, resulting in decreased intensity across entire spectrum or loss of intensity from a portion of the spectrum.
- **absorption spectrum**
 a graph that shows the percentage of each wavelength of light absorbed by a pigment.(e.g., chlorophyll).

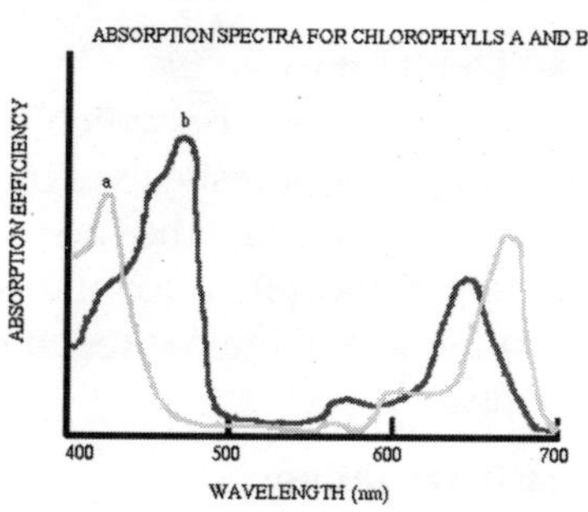

- **abundance**
 the estimated number of individuals of a species in an area or population.
- **acanthocarpous**
 a fruit showing prickles.
- **acarpous**
 describes a plant that is sterile.
- **acauline**
 not having a culm.
- **accession**
 a distinct sample of germplasm (cultivar, breeding line,

population) that is maintained in a gene bank for conservation and evaluation, in order to represent the genetic variation of a sample. Ideally 4,000 seeds are needed for genetically homogeneous lines and about 12,000 seeds for heterogeneous.

■ **accessory bud**
buds that are at or near the nodes but not in the axils of leaves.

■ **accidental host**
that type of host on which the pathogen or parasite lives only for a limited time, it has no particular importance for the reproduction of the pathogen or parasite.

■ **acclimatisation**
changes involving the synthesis of proteins, membranes and metabolites that occur in a plant in response to chilling or freezing temperatures that protect tissues or confer tolerance to the cold.

■ **acclimatised**
a state of physiological adjustment by plants to changed environmental or stress conditions.

■ **accumbent**
used to describe the first sprouts of an embryo when they lie against the body of the seed.

■ **accumulation centre**
an area where a great deal of variation of a given species or crop plant may be found, but which is not considered a centre of origin.

■ **acephalous**
not having a head.

■ **acephate**
a systemic insecticide that is used to control pests (e.g., aphids, scale and thrips).

■ **acerose**
needle like and stiff, like pine needles.

■ **acervate**
growth in heaps or groups.

■ **acetaldehyde**
a simple aldehyde that is a bridge product of alcoholic fermentation.

■ **achene**
a small, usually single-seeded, dry, indehiscent fruit formed

from a single carpel (e.g., the feathery achene of Clematis), variants of the achene include caryopsis, cypsela, nut and samara.

- **achiasmate**
  meiosis and/or chromosome pairing without crossing over and chiasma formation.

- **achlorophyllous**
  a plant or leaf without chlorophyll.

- **acicular leaf**
  a pointed or needle-shaped leaf (e.g., in conifers).

- **acid soil**
  specifically a soil with pH value < 7.0, which is caused by the presence of active hydrogen and/or aluminium ions, the pH value decreases as the activity of these ions increases.

- **acinaciform**
  shaped like a scimitar. (e.g., the shape of the pods of some beans).

- **acorn**
  the nonsplitting, one-seeded fruit of, for example, an oak tree.

- **acquired resistance**
  plant resistance to a disease activated after inoculation of the plant with certain micro-organisms or treatment with certain chemical compounds.

- **acrocarpic**
  fruits and/or seeds are formed on the top of a stem of a plant.

- **acropetal**
  toward the apex, the opposite of basipetal.

- **active collection**
  a collection of germplasm used for regeneration, multiplication, distribution, characterisation and evaluation. Ideally germplasm should be maintained in sufficient quantity to be available on request. It is commonly duplicated in a base collection and is often stored under medium to long-term storage conditions.

- **active immunity**
  all means and reactions that enable a plant to prevent an interaction with a pathogen.

- **active ingredient**
  in any pesticide product, the component that kills or controls target pest's active substance.

- **acyclic**
  not cyclic, in botany, an acyclic flower.

- **adaptable**
  capable of being adapted or able to adjust oneself readily to different environmental conditions.

- **adaptation**
  in the evolutionary sense, some heritable feature of an individual's phenotype that improves its chances of survival and reproduction in the existing environment.

- **adelphogamy**
  pollination involving a stigma and pollen belonging to two different individuals that are vegetatively derived from the same mother.

- **adosculation**
  the fertilisation of plants by pollen falling on the pistils.

- **adpressed**
  lying flat against (e.g., the rachilla against the palea in the grain of, e.g., barley and oats.)

- **adsorption complex**
  the various substances in the soil which are capable of adsorption (e.g., clay or humus).

- **adult resistance**
  resistance not expressed at the seedling stage, it increases with plant maturity.

- **advance crop**
  forecrop.

- **adventitious**
  growing from an unusual position (e. g., roots from a leaf or stem).

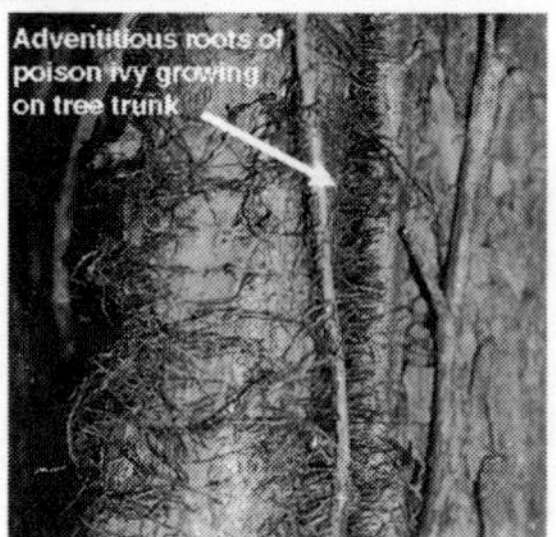

- **adventitious bud**
  a bud appearing in an unusual place (e.g., a bud on leaves).

- **adventitious plant**
  an individual that arises from somatic rather than reproductive tissue.

- **adventitious root**
  arising from any structure other than a root, e.g., from a node of a stem or from a leaf, the phenomenon is quite common in triticale and wheat, the adventitious root development is probably the response to accumulation of auxins in the base of the plant, sometimes this accumulation occurs

because the plants are waterlogged and the existing roots system is no longer a sink for the auxins due to lack of oxygen and/or dying off of the roots.

- **adventive**
  a plant that has been introduced but is not yet naturalised.

- **aeolian soil**
  a type of soil that is transported from one place to another by the wind.

- **aeration**
  bringing air into a substance, tissue or soil (e.g., by earthworms or digging and turning the soil to loosen).

- **aerator**
  any implement that is used for breaking up compacted soil to facilitate air and gas exchange.

- **aerenchyma**
  plant tissue containing large, intercellular air spaces.

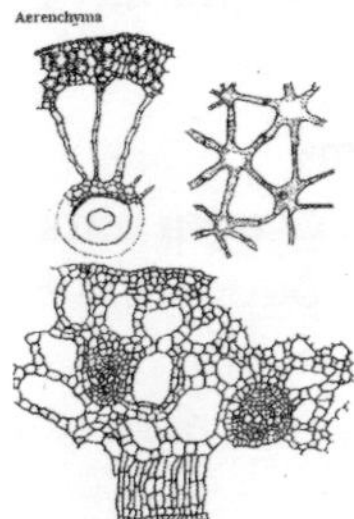

- **aerial pathogens**
  antagonistic micro-organisms that inhibit numerous fungal pathogens of aerial plant parts (e.g., *Tilletiopsis spp.* parasitise the cucumber powdery mildew fungus *Spaerotheca fuligena*), present in crop soils and exert a certain degree of biological control over one or many plant pathogens.

- **aerial root**
  in some epiphytic orchids, the leaves and the shoot axis are reduced or missing, then flatted and green roots take over the fixation of the plant and the function of the leaves (photosynthesis.)

- **aerial shoots**
  shoots growing high above the ground (e.g., trees, bushes, etc.).

- **afflux**
  the act of flowing to or toward some point or organ in a plant.

- **afforest**
  converting bare or cultivated land into forest.

- **afforestation**
  the establishment of forest by natural succession or by the planting of trees on land where they formerly did not grow.

- **after-ripening**
  a term for the collective changes that occur in a dormant seed that make it capable

of germination, it is usually considered to denote physiological changes.

- **agameon**
  a plant species reproducing exclusively by apomixes.

- **agamic**
  reproducing asexually.

- **agamic complex**
  it refers to hybrids or their derivatives that are partially or entirely reproduced by asexual seed formation.

- **agar**
  a complex polysaccharide obtained from certain types of seaweed (red algae), when it is heated with water and subsequently cooled to about + 45°C, it forms a gel.

- **aggregate fruit**
  a fruit development from several pistils in one flower, as in strawberry or blackberry.

- **agnesis**
  the absence of development.

- **agricultural biodiversity**
  the variety and variability of animals, plants and micro-organisms used directly or indirectly for food and agriculture (crops, livestock, forestry and fisheries). It comprises the diversity of genetic resources (varieties, breeds etc.) and species used for food, fuel, fodder, fibre and pharmaceuticals.

- **agriculture**
  the science of transforming sunlight energy into plant and animal products that can be utilised by humans the selective breeding of crops and farm animals has had an enormous impact on productivity in agriculture, modern varieties of crop plants have increased nutritional value and resistance to disease, recent developments in genetic engineering have enabled the potential use of transgenic organisms in agriculture to be explored.

- **agrobiology**
  the scientific study of plant life in relation to agriculture, especially with regard to plant genetics, cultivation and crop yield.

- **agroinfection**
  infection of plants via soilborne pathogens.

- **agronomy**
  the part of agriculture devoted to the production of crops and soil management, the scientific utilisation of agricultural land.

- **air layering**
  a method of plant propagation in which roots are induced to form around a stem, a very nar-

row strip of bark is removed from around the branch or stem, a sliver of wood can be inserted into the cut to keep it open, a bundle of moist sphagnum moss is tied securely around the cut area, the moss must remain moist and the roots of the plant

somewhat dry, new roots will sprout from the incision, the new plant is then cut off below the moss, potted and kept in a humid atmosphere until it is established.

■ **airlock**
an airtight chamber permitting passage to or from a space.

■ **albido**
the white tissue beneath the peel of citrus.

■ **albinism**
in plants, a deficiency of chromoplasts.

■ **albino**
a pigmentless white phenotype, determined by a mutation in a gene coding for a pigment-synthesising enzyme.

■ **albumen**
starchy and other nutritive material in a seed, stored as endosperm inside the embryo sac or as perisperm in the surrounding nucellar cells, in general, any deposit of nutritive material accompanying the embryo.

■ **albumin(e)**
any of certain proteins soluble in distilled water at neutral or slightly acid pH and in dilute aqueous salt solution. They coagulate by heat (e.g., leucosins in cereal grains, ricin in rice or legumelins in pulse seeds, which are mainly enzymes).

■ **albuminoid**
containing or resembling albumen or albumin.

■ **albuminous seed**
a seed having a well-developed endosperm or perisperm.

■ **alepidote**
having no scales of scurf, smooth.

■ **aleuron(e)**
a granulated protein that forms the outermost layer of a cereal grain, aleuron(e) grain.

■ **aleuron(e) layer**
a layer of cells below the testa of some seeds (e.g., cereals), which contains hydrolytic en-

zymes (e.g., amylases and proteases) for the digestion of the food stored in the endosperm, the production of enzymes is activated by gibberellins when the seed is soaked in water prior to germination.

- **alien addition line**
a line of plants with one or more extra chromosomes of an alien species.

- **alien chromosome**
a chromosome from a more or less related species transferred to a crop plant.

- **alien chromosome transfer**
cytogenetic methods that facilitate the transfer of individual chromosomes from one species to another.

- **alien gene transfer**
the transfer of genes between species or genera by different means.

- **alien germplasm**
genes introduced from a wild relative or nonadapted species of plants.

- **alien substitution line**
a line of plants in which one or more alien chromosomes from a certain donor species replace one or more chromosomes of a recipient species.

- **alkaline soil**
specifically a soil with pH value > 7.0 caused by the presence of carbonates of calcium, magnesium, potassium and sodium, commonly used for soils showing a pH value 8.5.

- **allelomorph**
a term that is commonly shortened to 'allele'.

- **alliaceous**
onionlike in smell or form.

- **allocompetition (intergenotype competition)**
cultivation at high plant density implies the presence of strong interplant competition. The individual plants, clones, lines or families are evaluated when being subjected to intergenotypic competition, also called intergenotype competition.

- **allogamous**
cross-fertilising in plants, as opposed to autogamous.

- **allogenetic**
cells or tissues related but sufficiently dissimilar in genotype to interact antigenically.

- **allometric**
growth in which the growth rate of one part of the plant differs from that of another part or of the rest of the plant.

- **allopatric**

applied to species that occupy separate habitats and that do not occur together in nature.

- **allopolyploid**

1. polyploid produced by the hybridisation of two species. See **amphidiploid**.

2. Plants with more than two sets of chromosomes that originate from two or more parents, the sets contain at least some nonhomologous chromosomes.

- **Allotetraploid, amphidiploid**

a plant that is diploid for two genomes, each from a different species

- **alluvial soil**

soils developed on fairly recent alluvium (a sediment deposited by streams and varying widely in particle size), usually they show no horizon development.

- **alternate**

not opposite to each other on the axis, but borne at regular intervals at different levels (e.g., of leaves.)

- **alternation of generation**

the alternation of two or more generations, reproducing themselves in different ways (i.e., alternation of gametophyte [sexual reproductive] and sporophyte [asexual reproductive] stages in the life cycle of a plant.)

- **aluminium (Al)**

has no specific importance in the metabolism of higher plants, small amounts of uptake favours the imbibition of the cytoplasm, higher concentrations in the soil may cause severe inhibition of plant growth. Aluminium tolerance is a main task of plant breeding in several regions of the world.

- **ambisexual**

a plant that has the reproductive organs of both sexes.

- **ambosexual**

ambisexual and more.

- **angiosperm**

1. plant whose seeds are enclosed within an ovary.

2. flowering plants.

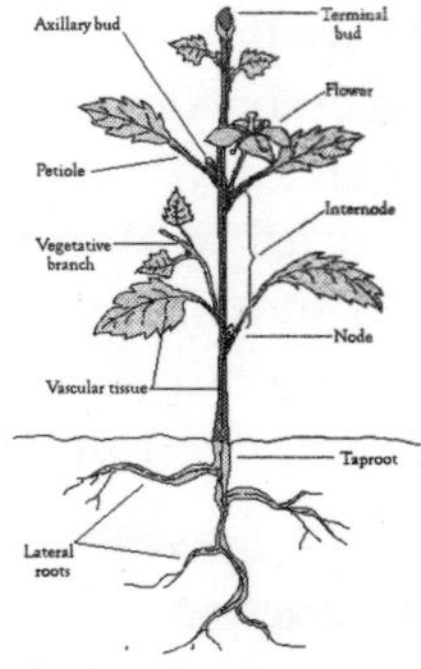

- **annealing**

spontaneous alignment of two complementary single polynucleotide (RNA or DNA or RNA and DNA) strands to form a double helix.

■ **antimutator mutation**
mutation of DNA polymerase that decreases the overall mutation rate.

■ **arrow**
the inflorescence of sugarcane.

■ **arsenic**
it may be found in plants in very low dosage, higher concentrations in soil may be toxic for plants.

■ **articulate**
jointed, having a node or joint.

■ **artificial light**
light other than sunlight, often from fluorescent tubes, used to grow plants in greenhouses or growing chambers usually out of season.

■ **artificial seeds**
based on embryogenic suspension cultures, embryos may be coated with water-soluble hygrogels and other substances in order to guarantee a proper germination even under field conditions.

■ **artificial selection**
plant selection by human or agronomic means.

■ **ascorbic acid**
the water-soluble vitamin C that occurs in large quantities in fruits and vegetables.

■ **asparagine (ASN)**
an amino acid found in storage proteins of plants (e.g., in peas and beans), the designation derived from the presence in asparagus.

■ **asparagus knife**
1. a tool for prying and pulling out long-rooted plants.
2. Dandelion weeder.
3. Fishtail weeder.

■ **aspirator**
an air-blast separator, a seed conditioning (cleaning) machine that uses air to separate according to specific gravity (weight) and resistance to air flow.

■ **assimilation**
the production of an organic substance from anorganic elements and compounds via photosynthesis carbon dioxide.

■ **assortative mating**
the mating of individuals with similar phenotypes.

■ **asynapsis**
chromosomes of meiosis I in which pairing either fails or is incomplete.

■ **attenuator stem**
a configuration of the leader transcript that signals transcription termination in attenuator-controlled amino acid operons.

- **autotroph**
  organism that is able to utilise carbon dioxide as a carbon source.

- **autozygosity**
  homozygosity in which the two alleles are identical by descent (i.e., they are copies of an ancestral gene).

- **B line**
  the fertile counterpart or maintainer line of an A line, does not have fertility restorer genes, used as the pollen parent to maintain the A line, used in hybrid seed production.

- **BAC vector**
  an *Escherichia coli* vector for DNA fragments, larger than cosmids, alternative to YAC vectors.

- **bacciferous**
  berry load-bearing, producing berries.

- **bacciform**
  berry-shaped.

- **backcross**
  a cross of an F1 hybrid or heterozygote with an individual of genotype identical to that on one or the other of the two parental individuals, matings involving a hybrid genotype are used in genetic analyses to determine linkage and crossing-over values.

- **backcross breeding**
  a system of breeding whereby recurrent backcrosses are made to one of the parents of a hybrid, accompanied by selection for a specific character(s).

- **backcrossing**
  backcross.

- **backfill**
  filling in a planting hole around roots with a soil mix for better establishing the plant.

- **backward selection**
  selection of parent plants based on results from a progeny test.

- **balanced design**
  an experimental design in which all treatment combinations have the same number of observations.

- **balanced lattice**
  a special group of balanced incomplete block, allows incomplete blocks to be combined into one or more separate complete replicates.

- **Balanced Tertiary Trisomic (BTT)**
  a specific interchange trisomic spontaneously selected or experimentally designed in a way that it is heterozygous, its trisomic progeny after selfing is genetically similar to the parent, the dominant allele is

present on the translocated chromosome linked to the break point. BTTs were used for hybrid seed production in barley.

- **balsam**

a mixture of resins and ethereal oils of sticky consistency, secreted by some plants.

- **band application**

the spreading of fertiliser or other chemicals over or next to, each row of plants in a field, as opposed to broadcast application.

- **band-seeding**

placing forage crop seed in rows directly above but not in contact with a band of fertiliser.

- **bark**

the outer skin of a tree trunk, outside the secondary, vascular cambium. It is composed of phloem tissue, which occurs as living inner and dead outer zones, the outer zone is penetrated by the cork layers formed from the cork cambia.

- **bark ringing**

a method used for forcing fruit trees to flower, a complete ring is cut around the trunk below the lowest branch and another ring is cut right below the first, the bark between the rings is removed, the scar should be covered with grafting wax.

- **basal node**

the node or joint at the base of the stem.

- **basal rosette**

in some plants, a cluster of leaves around the stem on or near the ground.

- **base**

a compound that reacts with an acid to give water (and a salt), a base that dissolves in water to produce hydroxide ions is called an alkali.

- **base collection**

a collection of germplasm that is kept for long-term, secure conservation and is not to be used as a routine distribution source, seeds are usually stored at subzero temperatures and low moisture content.

- **base seed**

particularly valuable seeds, usually derived from highly productive single plants (elite plants), which are used for seed production of commercially grown material, seed stock produced from breeder's seed by or under the control of, an appropriate agricultural authority, the source of certified seed, either directly or as registered seed.

- **basic form**
  primitive form.

- **basic seed**
  base seed.

- **bast**
  any of several strong, woody fibres, such as flax, hemp, ramie or jute, obtained from phloem tissue.

- **bast plant**
  crop plants used for fibre production, such as flax or hemp.

- **batch drying**
  drying seeds in relatively small quantities held in a stationary position (as opposed to drying in a continuous moving line.)

- **beak**
  the extension of the keel at the tip of the glume or lemma in wheat.

- **beat(ting) up**
  restocking failed areas in a crop or stand by further sowings or plantings. There are several other terms in use, e.g., 'blanking', 'filling', 'gapping', 'infilling', 'recruiting', 'reinforcement planting'. In forestry, it means to replace dead trees with new ones, especially during the early years of the establishment or re-establishment of a plantation.

- **bed**
  an area within a garden or lawn in which plants are grown.

- **bedding plant**
  a plant grown for its flowers or foliage that is suited by habit for growing in beds or masses.

- **beet**
  any of various biennial plants of the genus *Beta*, of the goosefoot family, especially *B. vulgaris*, having a fleshy red or white root and dark-green red-veined leaves sugarbeet is derived from *B. vulgaris* by selection for high sugar content.

- **behaviour flexibility**
  all means of plant behaviour permitting temporary adaptation to environmental conditions.

- **belowground biomass**
  biomass.

- **berry**
  a simple, fleshy or pulpy and usually many-seeded fruit that

has two or more compartments and does not burst open to release its seeds when ripe (e.g., banana, tomato, potato, grape).

- **berry-shaped**
  bacciform.

- **betacyanin**
  beet.

- **beta-galactosidase**
  the enzyme that splits lactose into glucose and galactose (coded by a gene (lac z) in the lac operon of *Escherichia coli*).

- **betaxanthin**
  beet.

- **bevel (of lemma)**
  a depression variable in depth in the base of the lemma, rounded in barley, transverse in oats.

- **biased**
  bias.

- **bi-cropping**
  a method of growing cereals in a leguminous living mulch, it could potentially reduce the need for synthetic inputs to cereal production while preventing losses of nutrients and increasing soil biological activity, also a method of low input production system for cereals.

- **biennial**
  a plant that lives for two years, during the first season food may be stored for the use during the flower and seed production in the second year.

- **biennial crop**
  biennial.

- **bifloral**
  showing two flowers.

- **bifoliate**
  showing two leaves.

- **bigerm**
  having two seeds.

- **bimodal distribution**
  a statistical distribution having two modes.

- **binemic**
  chromosomes that contain two DNA helices per metaphase chromatid.

- **biocatalyst**
  a biological substance used to cause a particular chemical or biochemical reaction.

- **biochemistry**
  the chemistry of life, the branch of chemistry that is concerned with biological processes.

■ **biocide**

a natural or synthetic substance toxic to living organisms.

■ **biocoenosis**

a community of organisms and its interaction with abiotic factors of habitat.

■ **biodiversity**

the existence of a wide variety of species (species diversity), other taxa of plants or other organisms in a natural environment or habitat or communities within a particular environment (ecological diversity) or of genetic variation within a species (genetic diversity). Genetic diversity provides resources for genetic resistance to pests and diseases. In agriculture, biodiversity is a production system characterised by the presence of multiple plant and/or animal species, as contrasted with the genetic specialisation of monoculture.

■ **biological containment**

precaution taken to prevent the spread of recombinant DNA molecules in the natural environment. Disabled host organisms (e.g., with s auxotrophic requirements or defective cell walls) together with non-transmissible cloning vectors are used. Biological containment is especially important when toxin genes from pathogens are expressed in *Escherichia coli* or other vectors.

■ **biological control**

the practice of using beneficial natural organisms to attack and control harmful plants, animal pests and weeds is called biological control or biocontrol. This can include introducing predators, parasites and disease organisms or releasing sterilised individuals. Biocontrol methods may be an alternative or complement to chemical and gene-engineered pest control methods.

■ **biological determinant**

a biological factor such as crop species, variety, weeds, insect pests or disease that determines the crop configuration and performance of a cropping pattern at a given site or area.

■ **biological pesticide**

a chemical which is derived from plants, fungi, bacteria or other natural synthesis and which can be used for pest control.

■ **biological yield**

the total yield of plant material (i.e., the total biomass including the economic yield, e.g., the grain yield). The larger the biological yield. The greater the photosynthetic efficiency.

- **biome**
  interactive groups of individuals of one or more species.

- **biopesticide**
  biological pesticide.

- **biopiracy**
  the collecting and patenting of plants and other biological material formerly held in common and their exploitation for profit.

- **bioreactor**
  a culture vessel used for experimental or large-scale bioprocessing.

- **bioseeds**
  seeds produced via genetic engineering of existing plants.

- **biotope**
  a portion of a habitat characterised by uniformity in climate and distribution of biotic and abiotic components.

- **biotrophy**
  obtains nutrients from living cells. biotrophs are typically obligate parasites. the term can also cover the phase in the infection process where a necrotroph does not destroy the host (hemibiotrophic).

- **bird pollination**
  ornithophily.

- **birimose**
  opening by two slits (e.g., anthers of plants).

- **bisexual**
  species comprises individuals of both sexes or a hermaphrodite organism in which an individual plant possesses both stamens and pistils in the flower

- **blade**
  the expanded portion of a leaf, petal or sepal.

- **blade joint**
  the flexible union between the leaf blade and the leaf sheath.

- **blanch**
  a method to whiten or prevent from becoming green by excluding light. blanching is applied to the stems or leaves of plants (e.g., celery, lettuce and endive). It is done either by banking up the soil around the stems, tying the leaves together to keep the inner ones from light or covering with pots, boxes, etc.

- **blasting**
  a plant symptom characterised by shedding of unopened buds, leads to a failure of producing fruits or seeds.

- **bleeding**
  exudation of the contents of the xylem stream at a cut surface due to root pressure.

■ **blend**
a term applied to mechanical seed mixtures of different crop varieties or species that have been mixed together to fulfil a specific agronomic purpose.

■ **blight**
a disease characterised by rapid and extensive death of plant foliage and applied to a wide range of unrelated plant diseases caused by fungi, when leaf damage is sudden and serious (e.g., fire blight of fruit trees, halo blight of beans, potato blight, etc.)

■ **blind**
without flowers, sterile.

■ **blindfold (trial)**
a trial to study soil heterogeneity (i.e., variation in the soil fertility). All plots contain the same genetically uniform plant material, the study may show that the growing conditions provided by a particular field may appear homogeneous when observed in some season and for some trait of a crop, but they may appear heterogeneous when observed in a different season or for some trait of a different crop, for a given crop. Different traits may differ with regard to their capacity to show soil heterogeneity.

■ **block**
a number of plots that offer the chance of equal growing conditions comparisons among the entries which are tested in the same block, offer unbiased estimates of genetic differences.

■ **blocking**
the procedure by which experimental units are grouped into homogeneous clusters in an attempt to improve the comparison of treatments by randomly allocating the treatments within each cluster or block.

■ **bloom**
the white powdery deposit often present on the surface of the stem, leaves and ears of cereals or sorghum, often of a waxy nature. In general, the flower of a plant or the state of blossoming.

■ **blooming**
in grasses, the period during which florets are open and anthers are extended.

■ **blossom**
the flower of a plant, especially of one producing an edible fruit, the state of flowering.

■ **blotch**
a disease characterised by large and irregularly shaped spots or blots on leaves, shoots and stem.

▪ **boll**

the fruit of cotton.

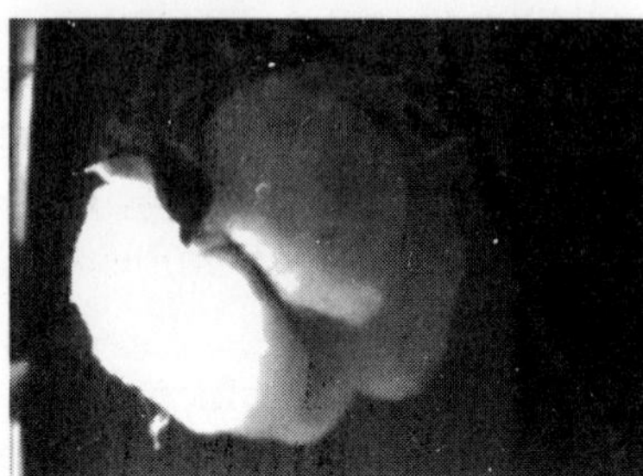

▪ **boll size**

weight in grams of seed cotton from one boll.

▪ **bolt**

formation of an elongated stem or seedstalk. In the case of biennial plants, this generally occurs during the second season of growth.

▪ **bolter**

they develop in long cold springs, with morning frosts and low temperatures not exceeding +5°C, causing vernalisation of the plants (e.g., in sugarbeet).

▪ **bolting**

production of seed stalks in the first season in a biennial crop (e.g., in beets.)

▪ **bonsai**

a tree or shrub grown in a container or special pot and dwarfed by pruning, pinching and wiring to produce a desired shape.

▪ **boot**

the lower part of a cereal plant.

▪ **boot(ing) stage**

it refers to the growth stage of grasses at the time the head is enclosed by the sheath of the uppermost leaf.

▪ **border effect**

the environmental effect on plots that are on the edge of an experimental area.

▪ **boron (B)**

a non-metallic element occurring naturally only in combination, as in borax or boric acid. Boron can cause toxicity in several crop plants, as a micronutrient deficiency of boron can be as severe.

▪ **botanical pesticides**

pesticides whose active ingredients are plant-produced chemicals such as nicotine, rotenone or strychnine.

▪ **botany**

the science of plants, the branch of biology that deals with plant life, the plant life of a region, the biological characteristics of a plant or group of plants.

▪ **botuliform**

cylindrical with rounded ends, sausagelike in form.

- **bough**
  the main arm or branch of a (fruit) tree.

- **brachyomeiosis**
  an abnormal meiosis characterised by omission of the second meiotic division.

- **bract**
  a modified leaflike structure occurring in the inflorescence.

- **bracteole**
  a little bract borne on the flowerstalk above the bract and below the calyx.

- **branch**
  an axillary (lateral) shoot or root.

- **brassinosteroid**
  brassinosteroids are endogenous, plant-growth-promoting natural products with structural similarities to animal steroid hormones. They affect cell elongation and proliferation. Distinct from auxins, cytokinins and gibberellic acids, although they interact with them.

- **breed**
  an artificial mating group derived from a common ancestor or for genetic analysis, in breeding, a line having the character type and qualities of its origin. In general, a group of plants, developed by humans that will not keep their characteristics in the wild.

- **breeder**
  a person who raises plants primarily for breeding purposes.

- **breeder('s) seed**
  seed or vegetative propagating material increased by the originating or sponsoring plant breeder or institution. It represents the true pedigree of the variety, it is used for the production of genetically pure foundation, registered and certified seeds.

- **breeding**
  the propagation and genetic modification of organisms for the purpose of selecting improved offspring. Several techniques of hybridisation and selection are applied.

- **breeding cycle**
  the shortest period between successive generations from germination of a seed to reproduction of the progeny, i.e., the seed-to-seed cycle.

- **breeding line**
  a group of plants with similar traits that have been selected for their special combination of traits from hybrid or other populations, it may be released or used for further breeding approaches.

- **breeding orchard**
  a planting of selected trees, usually clonally or grafted propagated, it is designed to ease breeding work.

- **breeding population**
  a group of individuals selected from a wild, experimental or crossing population for use in a breeding programme, usually phenotypically selected for desirable traits.

- **breeding size**
  the number of individuals in a population involved in reproduction during particular generations and breeding procedures.

- **breeding strategy**
  prescription for breeding, a sound breeding strategy. Searches for an optimal compromise between genetic gain, gene diversity, cost, time and other factors.

- **breeding system**
  the system by which a species reproduces, more specifically, the organisation of mating that determines the degree of similarity and/or difference between gametes effective in fertilisation.

- **bridging cross**
  a cross made to transfer alleles between two sexually isolated species by first transferring the alleles to an intermediate species that is sexually compatible with both.

- **bridging species**
  a species used in a bridging cross in order to bring together the two incompatible species.

- **broad-base terrace**
  a low embankment that is constructed across a slope to reduce runoff and/or erosion (e.g., in rice or grape cultivation).

- **broadcast**
  scattered upon the ground with the hand (e.g., in sowing seed, instead of sowing in drills or rows).

- **broadleaf**
  sometimes used to designate a broad group of non-grasslike (weedy) plants.

- **C3 pathway**
  most common pathway of carbon fixation in plants, this photosynthesis produces at first a 3-carbon (C_3) compound (phosphoglyceric acid), in C_3 plants, about 25 percent of the net carbon uptake is re-evolved immediately in photorespiration.

- **C4 pathway**
  a carbon fixation found in some plants that have high rates of

growth and photosynthesis and that are adapted to high temperatures, strong light, low carbon dioxide levels and low water supply. This photosynthesis produces at first a 4-carbon (C_4) compound (phosphoenolpyruvate, PEP). In C_4 plants, photorespiration is suppressed to a very large extent due to the presence of a very efficient C_2-concentrating mechanism.

- **callus**
an undifferentiated clone of plant cells.

- **carbon source**
a nutrient (such as sugar) that provides carbon skeletons needed for synthesis of new organic molecules (anabolism).

- **catabolite repression**
repression (inactivation) of certain sugar-metabolising operons (e.g., lac) in favour of glucose utilisation when glucose is the predominant carbon source in the environment of the cell.

- **catalyst**
a substance that increases the rate of a chemical reaction without itself being permanently changed.

- **cell cycle**
the cycle of cell growth, replication of the genetic material and nuclear and cytoplasmic division.

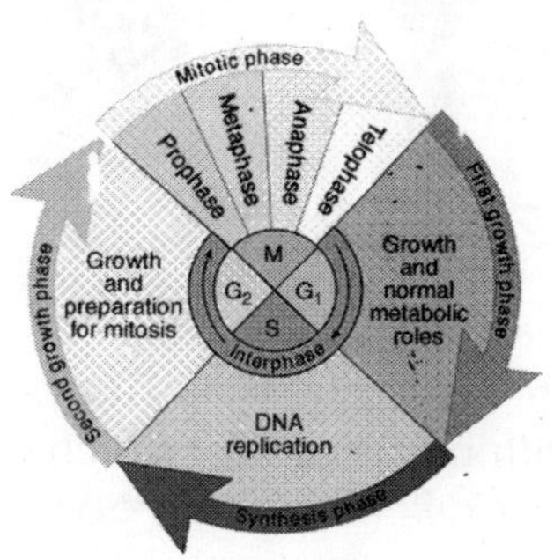

- **cell division**
the process by which two cells are formed from one. See **meiosis** and **mitosis.**

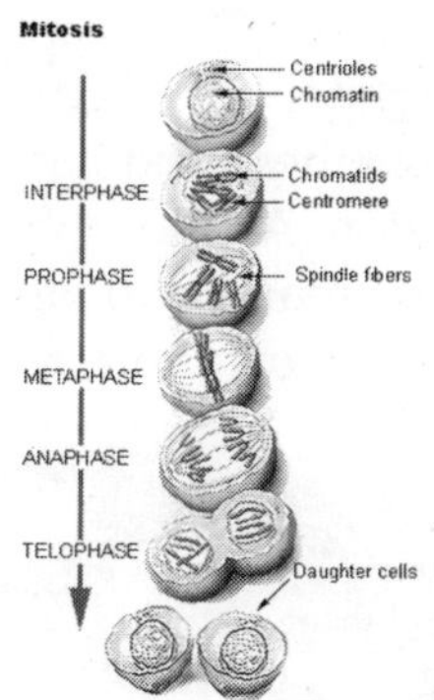

- **cell lineage**
a pedigree of cells related through asexual division.

- **centromere**
a kinetochore, the constricted region of a nuclear chromosome, to which the spindle fibres attach during division.

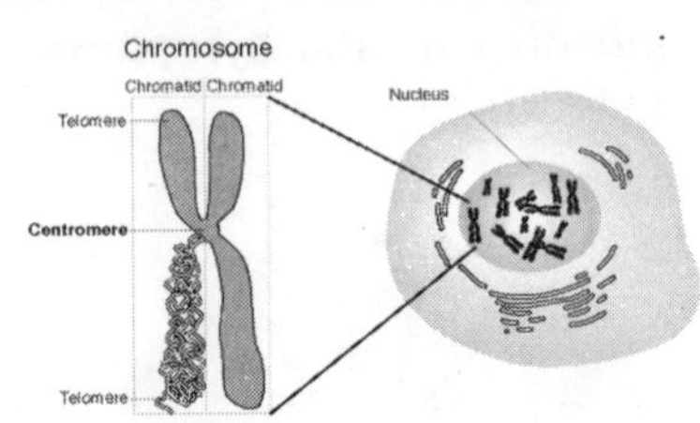

- **character**
  an attribute of individuals within a species for which heritable differences can be defined.

- **character difference**
  alternative forms of the same attribute within a species.

- **chloroplast**
  the organelle that carries out photosynthesis and starch grain formation. A chlorophyll-containing organelle in plants that is the site of photosynthesis.

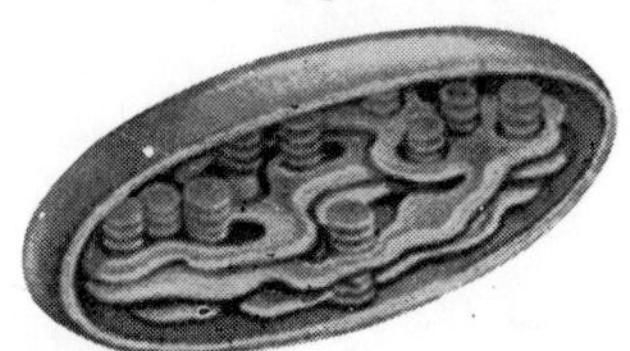

- **chromatid**
  one of the two side by side replicas produced by chromosome replication in mitosis or meiosis. Subunit of a chromosome after replication and prior to anaphase of meiosis II or mitosis. At anaphase of meiosis II or mitosis when the centromeres divide and the sister chromatids separate each chromatid becomes a chromosome.

Appearance of Homologous Pair of Chromosomes at Prophase 1

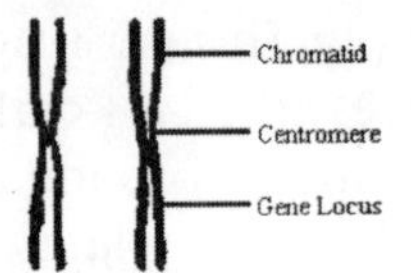

- **chromomere**
  a small beadlike structure visible on a chromosome during prophase of meiosis and mitosis. Dark regions of chromatin condensation in eukaryotic chromosomes at meiosis or mitosis.

- **chromosome**
  a linear end-to-end arrangement of genes and other DNA, sometimes with associated protein and RNA. The form of the genetic material in viruses and cells. A circle of DNA in prokaryotes, a DNA or a RNA molecule in viruses, a linear nucleoprotein complex in eukaryotes.

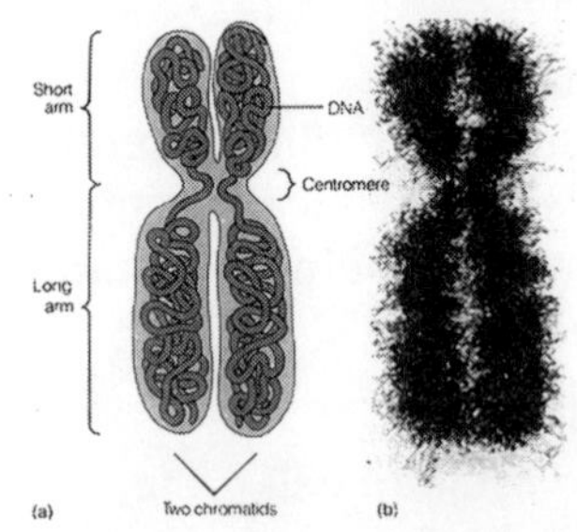

- **chromosome aberration**
any type of change in the chromosome structure or number.

- **chromosome banding**
a technique for staining chromosomes so that bands appear in a unique pattern particular to the chromosome.

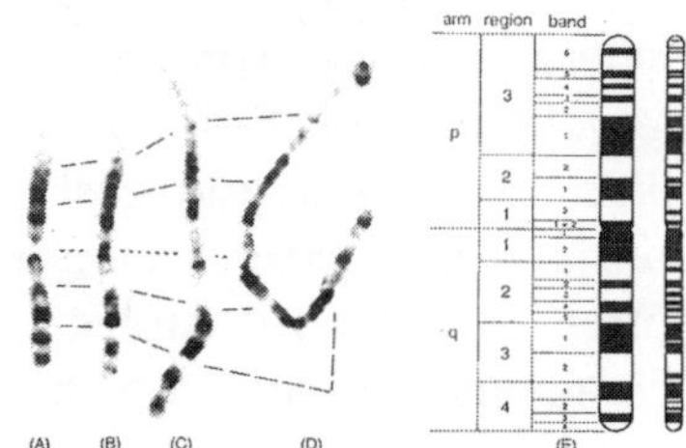

- **chromosome jumping**
a technique of isolating clones from a genomic library that are not contiguous by skipping a region between known points on the chromosome. Done usually to bypass regions that are difficult or impossible to walk through or regions known not to be of interest.

- **chromosome loss**
failure of a chromosome to become incorporated into a daughter nucleus at cell division.

- **chromosome map**
see **linkage map**.

- **chromosome puff**
a swelling at a site along the length of a polytene chromosome, the site of active transcription. A diffuse uncoiled region in a polytene chromosome where transcription is actively taking place.

Puff in Polytene Chromosome

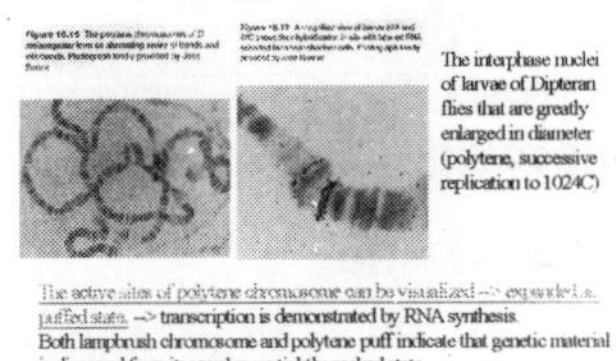

- **chromosome rearrangement**
a chromosome mutation involving new juxtapositions of chromosome parts.

- **chromosome set**
the group of different chromosomes that carries the basic set of genetic information for a particular species.

- **chromosome theory of inheritance**
the theory that chromosomes are linear sequences of genes. The unifying theory stating that inheritance patterns may be generally explained by assuming that genes are located in specific sites on chromosomes.

- **circular permutation**
some phages with linear genomes (e.g., T2) are both terminally redundant and circularly permuted. If the genetic information is represented by

ABCDEFGH
then a circular permutation would generate molecules
ABCDEFGH
BCDEFGHA
CDEFGHAB
DEFGHABC
EFGHABCD
and so on.

- **clinal selection**
  selection that changes gradually along a geographic gradient.

- **clone**
  genetically engineered replicas of DNA sequences.

- **clone bank**
  see **genomic library**.

- **cloning**
  the process of asexually producing a group of cells (clones), all genetically identical, from a single ancestor. In recombinant DNA technology, the use of DNA manipulation procedures to produce multiple copies of a single or segment of DNA is referred to as cloning DNA.

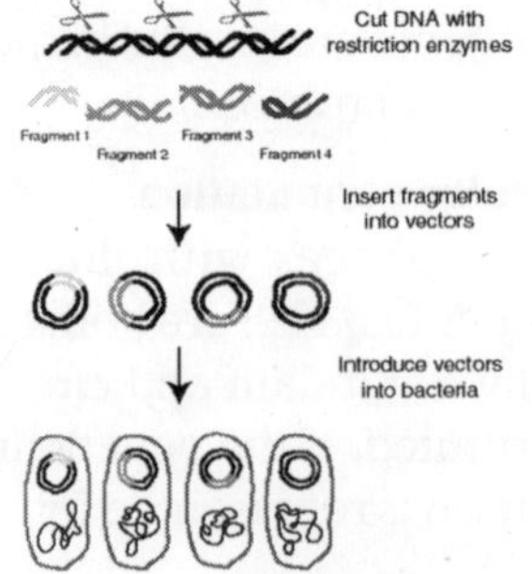

- **cloning vector**
  DNA molecule originating from a virus, a plasmid or the cell of a higher organism into which another DNA fragment of appropriate size can be integrated without loss of the vectors capacity for self- replication. Vectors introduce foreign DNA into host cells, where it can be reproduced in large quantities. Examples are *plasmids, cosmids* and *yeast artificial chromosomes*. Vectors are often *recombinant* molecules containing DNA sequences from several sources.

- **cm(centimorgan)**
  see **map unit**.

- **coccus**
  a spherical bacterium.

- **code**
  see **genetic code**.

- **coding strand**
  the DNA strand with the same sequence as the transcribed mRNA (given U in RNA and T in DNA) and containing the linear array of codons which interact with anticodons of tRNA during translation to give the primary sequence of a protein. Compare with **anticoding strand**.

- **codon**
  a section of DNA (three nucleotide pairs in length) or RNA

(three nucleotides in length) that codes for a single amino acid. A sequence of three RNA or DNA nucleotides that specifies (codes for) either an amino acid or the termination of translation.

- **codon preference**
  the idea that for amino acids with several codons one or a few are preferred and are used disproportionately. They would correspond with tRNAs that are abundant.

- **coefficient of coincidence**
  in a three point cross the number of observed double crossovers divided by the number expected based on the observed occurrence of single crossovers. The ratio of the observed number of double recombinants to the expected number.

- **coefficient of relationship (r)**
  the proportion of alleles held in common by two related individuals.

- **cohesive end**
  a single-stranded end to a linear duplex DNA molecule, which can hydrogen-bond with a complementary single-strand base sequence from the end of the same or another DNA molecule.

- **cointegrate**
  a fusion of two elements. An intermediate structure in replicative transposition. The product of the fusion of two circular elements to form a single, larger circle.

- **colinearity**
  the one to one linear correspondence between the order of codons in a coding sequence and the order of amino acids in the protein encoded. A linear map of mutation sites within a gene corresponds to the linear location of amino acid substitutions within the polypeptide encoded by that gene.

- **colony**
  a visible clone of cells.

- **common ancestry**
  the state of two individuals when they are blood relatives. When two parents have a common ancestor their offspring will be inbred.

- **compartment**
  the existence of boundaries within the organism beyond which a specific clone of cells will never extend during development.

- **competence factor**
  a surface protein that binds extracellular DNA and enables the cell to be transformed.

- **competent**
able to take up exogenous DNA and thereby be transformed.

- **Complementary DNA (CDNA)**
synthetic DNA reverse transcribed from a specific RNA through the action of the enzyme reverse transcriptase. DNA synthesised by reverse transcriptase using RNA as a template.

- **complementary sequences**
nucleic acid base sequences that can form a double-stranded structure by matching base pairs. The complementary sequence to G- T- A- C is C- A- T-G.

- **complementation**
the production of a wild-type phenotype when two different mutations are combined in a diploid or a heterokaryon. The production of the wild-type phenotype by a cell or an organism that contains two mutant genes. If complementation occurs the mutations are almost certainly non-allelic (i.e. in different genes).

- **complementation group**
a group of mutant genes which do not complement each other. A cistron (determined by the cis-trans complementation test).

- **complementation test**
a mating test to determine whether two different recessive mutations (a1,a2) on opposite chromosomes (trans, a1+/+a2) of a diploid or partial diploid will not complement (i.e., have a mutant phenotype) each other, but the same two recessive mutations on the same chromosome (cis, a1a2/++) in a diploid or partial diploid show a wild-type phenotype, a test for allelism. A test to determine whether two mutant sites are in the same functional unit or gene.

- **complete flower**
a flower that has pistils, stamens, petals and sepals.

- **complete linkage**
the state in which two loci are so close together that alleles of these loci are virtually never separated by crossing over.

- **complete medium**
a culture medium that is enriched to contain all of the growth requirements of a strain of organisms.

- **complete penetrance**
the situation in which a dominant gene always produces a phenotypic effect or a recessive gene in the homozygous state always produces a detec effect.

- **complex heterozygous**
special type of genetic system based on the heterozygosity for multiple reciprocal translocations.

- **complex locus**
a cluster of two or more closely linked and functionally related genes constituting a pseudoallelic series.

- **component of fitness**
a particular aspect in the life cycle of an organism upon which natural selection acts.

- **component of variance**
variance.

- **composite**
a plant of the immense family *Compositae*, regarded as comprising the most highly developed flowering plants , a mixture of genotypes from several sources, maintained by normal pollination.

- **composite cross**
a population derived from the hybridisation of several parents, either by hand-pollination or by the use of male sterility.

- **composite fruit**
a seed distribution unit that includes many ovaries connected by fruit walls or other sui tissue, if the flower basis (receptacle) or other flower components are thick and fleshy (e.g., in strawberry, apple or fig), it is called false fruit or pseuduocarp.

- **composite mixture**
breeder seed obtained by mechanically combining seed from two or more strains. The mixture is increased through successive steps in a certified seed programme and distributed as a synthetic variety.

- **composite transposon**
a transposon constructed of two IS (insertion sequence) elements flanking a control region that frequently contains host genes.

- **compost**
plant and animal residues that are arranged into piles and allowed to decompose.

- **compound cross**
a combination of desirable genes from more than two inbred lines, breeding strains or varieties.

- **compound cyme**
a determinate inflorescence where there is secondary branching and each ultimate unit becomes a simple cyme.

- **concatemer**
tandem repeats of identical DNA molecules, lamda phage DNA must be concatemer in

order to be packaged lamda phage.

- **concatenate**
interlocked circles (e.g., plasmids).

- **concave**
shaped like the inside of an egg.

- **concordance**
the amount of similarity in phenotype between individuals.

- **condensation of chromosomes**
chromosome contraction coiling.

- **conditional mutation**
a mutation that has the wild-type phenotype under certain (permissive) environmental conditions and a mutant phenotype under other (restrictive) conditions.

- **conditional-lethal mutation**
a mutation that is lethal under one condition but not lethal under another condition.

- **conditioned dominance**
dominance affected by the presence of other genes or by environmental influence.

- **conditioned storage**
storage of seed under controlled conditions of temperature and relative humidity.

- **conditioner**
a material or substance added to a fertiliser that keeps it flowing free.

- **conditioning**
the term used to describe the process of cleaning seed and preparing it for market, sometimes called processing

- **conduction**
plasmid mobilisation involving cointegrate formation.

- **conductivity test**
an electrical conductivity test that associates the concentration of leachates from seeds, after soaking in water, to their quality.

- **cone**
a fruit with overlapping scales in which seeds are formed.

- **cone collection**
harvesting of cones after seed maturation but before their dispersal.

- **confidence belt**
confidence limit.

- **confidence limit**
a statistical term for a pair of numbers that predict the range of values within which a particular parameter lies for a given level of confidence (probability.)

- **congenic strain**
  a variant plant strain that is obtained by backcrossing a donor plant strain to an inbred parental strain for at least eight generations while maintaining by appropriate selection the presence of a small genetic region derived from the donor strain.

- **conical divider**
  an inverted metal cone below a spout from a hopper, the seeds fall over the cone to be evenly dispersed. A series of bugle or riffle dividers separate the seeds into channels.

- **conidiophore**
  a threadlike stalk upon which conidia (spores) are produced, a specialised hypha upon which one or more conidia may bear.

- **conidium**
  any asexual spore formed on a conidiophore.

- **conifer**
  a species of plant that bears it seeds in cones, such as a pine tree.

- **coniferous tree**
  conifer.

- **conjugation**
  a process whereby organisms of identical species, but opposite mating types, pair and exchange genetic material (DNA). In molecular biology, natural process of DNA transfer between bacteria in which the DNA is never exposed. It is insensitive to externally added DNase.

- **conjugation tube**
  see **pilus**.

- **connective**
  the tissue joining the two cells of an anther.

- **conoidal**
  nearly conical.

- **consanguineous**
  meaning between blood relatives , usually refers to inbreeding or incestuous matings.

- **consanguinity**
  genetic relationship. Consanguineous individuals have at least one common ancestor in the preceding few generations.

- **consensus sequence**
  if a particular nucleotide sequence is always found with only minor variations, then the usual form of that sequence is called consensus sequence. The term is also used for genes that encode the same protein in different organisms.

■ **conservation**
maintenance of environmental quality and resources.

■ **conservation tillage**
seed bed preparation systems that have about 30 percent or more of the residue cover on the surface after planting.

■ **conservative change**
an amino acid change that does not affect significantly the function of the protein.

■ **conserved sequence**
a base sequence in a DNA molecule (or an amino acid sequence in a protein) that has remained essentially unchanged throughout evolution.

■ **constant region**
a region of an antibody molecule that is nearly identical with the corresponding regions of antibodies of different specificities.

■ **constitutive**
always expressed in an unregulated fashion (when referring to gene control.)

■ **constitutive heterochromatin**
heterochromatin that surrounds the centromere. Specific regions of heterochromatin are always present and in both homologues of a chromosome. See **satellite DNA.**

■ **constitutive heterochromatin**
the material basis of chromosomes or segments that exhibit heterochromatic properties under most conditions (e.g., centromeric or telomeric heterochromatin.)

■ **constitutive mutation**
causes genes that usually are regulated to be expressed without regulation.

■ **constriction**
an unspiralised segment of fixed position in the metaphase chromosomes (nucleolar ~ , primary or centric ~ , secondary ~).

■ **containment**
measures taken to prevent release of recombinant DNA molecules into the natural environment, biological and physical methods are applied.

■ **contig map**
a map depicting the relative order of a linked library of small overlapping clones representing a complete chromosomal segment.

■ **contigs**
groups of *clones* representing overlapping regions of a genome.

- **contiguous (contig) map**
the alignment of sequence data from large, adjacent regions of the genome to produce a continuous nucleotide sequence across a chromosomal region.

- **contiguous genes**
genes physically close on a chromosome that when acting together express a phenotype.

- **continuous culture**
an in vitro suspension culture continuously supplied with nutrients by the inflow of fresh medium, the culture volume is normally constant.

- **continuous replication**
the uninterrupted replication of DNA in the 5' to 3' direction using a 3' to 5' template.

- **continuous scale**
a scale for scoring quantitative data for which the number of potential values is not predefined and is potentially limitless (e.g., seed weight in grams).

- **continuous variation**
variation measured on a continuum rather than in discrete units or categories (e.g., height in human beings).

- **contrasting genetic character**
a character with marked phenotypic differences.

- **control**
an economic reduction of crop losses caused by plant diseases.

- **control pollination**
in horticulture and forestry, to purposely pollinate the female flowers of a tree with pollen from a known source. Usually the flowers are isolated from undesirable pollen by covering them with a pollen-tight cloth or paper bag before they are receptive. It is a way to produce full-sib families.

- **controlling element**
a term used by maize geneticists to indicate a mobile genetic element capable of producing an unstable mutant target gene: two types exist, the regulator and the receptor elements.

- **controlling gene**
a gene that is involved in turning on or off the transcription of structural genes. Two types of genetic elements exist in this process, a regulator and a receptor element the receptor elements is one that can be inserted into a gene, making it a mutant and can also exit from the gene both of these functions are under control of the regulator element.

- **convergence breeding**
a breeding method involving the reciprocal addition to each

of two inbred lines of the dominant favourable genes lacking in one line and present in the other. Backcrossing and selection are performed in parallel, each of the original lines serving as the recurrent parent in one series.

- **convergence-divergence selection**

a breeding scheme in which selection of promising genotypes is made in a bulk population at different locations followed by massing of selection and allowing mating among them in a pollination field. The harvested bulk seeds constitute the basis for the next propagation cycle.

- **convex**

shaped like the outside of an egg.

- **co-orientation**

centromere orientation.

- **cop**

coefficient of parentage.

- **copper (Cu)**

a malleable ductile metallic element having a characteristic reddish brown colour. As a trace element it is needed by plants, deficiency can cause severe problems of growth. Like iron efficiency, copper efficiency is genetically controlled (e.g., on rye chromosome arm 5RL a dominant gene and/or gene complex is located, increasing Cu efficiency not only in rye but also in wheat when the gene is transferred into the recipient).

- **copper fist**

configuration of a DNA-binding protein that resembles a fist closed around a penny. In this case the penny is copper ions, the knuckles of the fist of the yeast ACE1 protein interact with the promoter of the metallothionein gene, enhancing its transcription.

- **coppice**

natural regeneration originating from stump sprouts, stool shoots or root suckers.

- **coppice method**

a method of regenerating a forest stand in which the cut trees produce sprouts, suckers or shoots.

- **coppice selection method**

a method in which only selected shoots of usable size are cut at each felling, leading to uneven-aged stands.

- **coppice shoot**

any shoot arising from an adventitious or dormant bud near the base of a woody plant that has been cut back.

- **coppice-of-two-rotations method**
a coppice method in which some of the coppice shoots are reserved for the whole of the next rotation, the rest being cut.

- **coppice-with-standards method**
regenerating a forest stand by coppicing. Selected trees grown from seed are left to grow to larger size than the coppice beneath them. The method is used to provide seeds for natural regeneration of standards in subsequent rotations.

- **copulation**
the fusion of sexual elements.

- **copy error**
an error in the DNA replication process giving rise to a gene mutation.

- **copy gene**
genetic material incorporating the genetic code for a desirable trait which has been copied from DNA of the donor to the host organism.

- **copy-choice hypothesis**
an incorrect hypothesis that stated that recombination resulted from the switching of the DNA-replicating enzyme from one DNA homologue to the other.

- **copy-choice model**
a model of the mechanism for crossing over, suggesting that crossing over occurs during chromosome division and can occur only between two supposedly new nonsister chromatids. The experimental evidence does not support this model.

- **cordon**
an extension of the grapevine trunk, usually horizontally oriented and trained along the trellis wires, it is considered permanent (or perennial) wood.

- **core collection**
the basic sample of a germplasm collection, it is designed to represent the wide range of diversity in terms of morphology, geographic range or genes. It contains, with a minimum of repetitiveness, the genetic diversity of a crop species and its wild relatives. It is not intended to replace existing gene banks collections but to include the total range of genetic variation of a crop in a relatively small and manageable set of germplasm accessions.

- **corepressor**
the metabolite that when bound to the repressor (of a repressible operon) forms a functional unit that can bind to its operator and block transcription.

- **coriaceous**
leathery.

- **cork**
in woody plants, a layer of protective tissue that forms below the epidermis.

- **cork cambium**
phellogen.

- **cork layer**
layer of dead protective tissue between the bark and cambium in woody plants

- **corm(us)**
an underground storage organ formed from a swollen stem base, bearing adventitious roots and scale leaves. It may function as an organ of vegetative reproduction or in perennation.

- **corn**
the edible seed of cereal plants other than maize caryopsis.

- **corneous**
it refers to hard, vitreous or horny endosperm in cereal grains.

- **corn-loft**
granary.

- **corolla**
a collective term for all the petals of a flower, a non-reproductive structure, often arranged in a whorl. Encloses the reproductive organs.

- **correction**
the production (possibly by excision and repair) of a properly paired nucleotide pair from a sequence of hybrid DNA that contains a mismatched base pair. See **mismatch repair**.

- **correlation**
the degree to which statistical variables vary together. Measured by the correlation coefficient, which has a value from zero (no correlation) to -1 or $+1$ (perfect negative or positive correlation).

- **correlation breaker**
outlier.

- **corresponding gene pair**
a pair of genes in a parasite that corresponds with a pair of genes in a host, which function together to bring about a specific outcome cortex rind.

- **corymb**
a racemose inflorescence in which the lower pedicels are longer than the upper so that the flower lies as a dome or dish

and the outline is roundish or flattish.

- **cos site**
  the site of the circular form of phage lamda or others that is cleaved by the terminase to generate the cohesive 12 bp 5' overhang ends of the linear phage as it is packaged into the capsid cosmid lamda phage.

- **cosegregation**
  the tendency for closely linked genes and genetic markers to segregate (be inherited) together.

- **cosmid**
  a hybrid plasmid that contains cos sites at each end. Cos sites are recognised during head filling of lambda phages. Cosmids are useful for cloning large segments of foreign DNA (up to 50 kb).

- **cosmopolite**
  plant of worldwide distribution.

- **cosuppression**
  silencing of a gene by addition of transgenic DNA copies or infection by a virus. This term, which can refer to silencing at the post-transcriptional (PTGS) or transcriptional (TGS) level, has been primarily adopted by researchers working with plants post-transcriptional silencing.

- **cotransduction**
  the simultaneous transduction of two or more genes. The simultaneous transduction of two bacterial marker genes.

- **cotransformation**
  the simultaneous transformation of two bacterial marker genes.

- **cotyledon**
  the leaf-forming part of the embryo in a seed. It may function as a storage organ from which the seedling draws food or it may absorb and pass on to the seedling nutrients stored in the endosperm. Once it is exposed to light it develops chlorophyll and functions photosynthetically as the first leaf.

- **cotyledonary node**
  the point of attachment of the cotyledons to the embryonic axis.

- **coulter**
  a sharp blade or wheel attached to the beam of a plough, used to cut the ground in advance of the ploughshare.

- **coumarin**
  a white crystalline compound ($C_9H_6O_2$) with a vanilla-like odour. It gives sweetclover its distinctive odour. It is also known as a chemical growth

inhibitor that has germination-inhibiting capability.

- **couple method (of breeding)**
  a breeding method exclusively used in breeding of allogamous plants, from an original population (e.g., of sugarbeet), single plants are selected and, subsequently, pair-wise crossed, preventing unwished pollination. The crossing partners should be as similar as possible in spite of colour, growth habit, etc. The offspring is grown in separate plots during the following year. The selection of individuals from the plots and a repeated pair-wise crossing can be realised during the fourth year. During the fifth year offspring is grown in plots and again selected for progeny testing.

- **coupling**
  arrangement of wild-type and mutant alleles at two linked loci in which both mutants are on the same chromosome and both wild-type alleles on the homologue (ab/AB). See **repulsion**.

- **coupling conformation**
  linked heterozygous gene pairs in the arrangement, AB/ab.

- **coupling of factors**
  linkage in which both dominant alleles are in the one parent linkage.

- **cover crop**
  a crop grown between orchard trees or on fields between the cropping seasons of a main crop, to protect the soil against erosion and leaching and for improvement of soil.

- **CPDNA**
  Chloroplast DNA.

- **crease**
  the fold on a cereal grain.

- **criss-cross inheritance**
  the transmission of a gene from mother to son or father to daughter criss-crossing.

- **criss-crossing**
  a continuous, rotational cross-breeding system alternately using males or pollinators of two different breeds. This system is simple to manage and breeds its own replacements. It utilises the benefit of hybrid vigour. Compared to the common F1, some hybrid vigour can be lost, but that loss is more than compensated for by reduced management effort and cost.

- **cristae**
  mitochondrion.

- **critical difference**
  a value indicating least significant difference at values greater than which all the differences are significant.

- **CRNA**
  see **complementary RNA**.
- **crop**
  a species expressly cultivated for use as crop plant.
- **crop divider (at the harvester)**
  separates the standing crop from the material being cut.
- **crop evolution**
  the adaptation of a crop over generations of association with humans.
- **crop plant**
  a plant expressly cultivated for use. The majority of crops can be classified as (1) root and tuber crops (potato, yams), (2) cereals (e.g., wheat, oats, barley, rye, rice, maize), (3) oil and protein crops (rapeseed, pulses), (4) sugar crops (sugarbeet, sugarcane), (5) fibre crops (cotton, jute) or (6) forage crops (grasses, legumes) agronomic crops can be classified as (a) green manure crops, (b) cover crops, (c) silage crops or (d) companion crops. About 2 percent of the 250,000 higher plant species are used in agriculture, horticulture, etc. Economically, the most important families are the legumes and the grasses, which account for more than a quarter of the total species. They are followed by Rosaceae, Compositae, Euphorbiaceae, Labiatae and Solanaceae, all with more than 100 taxa, among the families with 50 to 100 crop species. Liliaceae, Agavaceae and Palmae are worth mentioning, whereas more than 50 percent of the families have fewer than ten crop species.
- **crop residue**
  that portion of a plant left in the field after harvest (maize stalk or stover, stubble.)
- **crop rotation**
  the alternation of the crop species grown on a field, usually this is done to reduce the pest and pathogen population or to prevent one-track exhaustion.
- **cross**
  bringing together of genetic material from different individuals in order to achieve genetic recombination.
- **cross back**
  backcrossing.
- **cross coancestry**
  refers to the average of the elements in a coancestry matrix excluding the self-coancestry on the diagonal, thus the expected inbreeding following random mating in a population without inbreeding.

- **cross(ing)-over unit**
  a 1 percent crossing-over value between a pair of linked genes, MORGAN unit.
- **crossability**
  the ability of two individuals, species or populations to cross or hybridise.
- **crossbred**
  self.
- **crossbreed**
  fertilisation between separate individuals.
- **crossbreeding**
  outbreeding or the breeding of genetically unrelated individuals, this may entail the transfer of pollen from one individual to the stigma of another of a different genotype.
- **crossbreeding barrier**
  a pre- and/or post fertilisation condition (i.e., progamous or postgamous incompatibility) that prevents or reduces crossbreeding or any form of gene transfer. It is caused by genetic, environmental, physical or chemical influences.
- **cross-fertilisation**
  the fusion of male and female gametes from different genotypes or individuals of the same species, as base of genetic recombination allogamy cross-pollination.
- **crossing barrier**
  any of the genetically controlled mechanisms that either entirely prevent or at least significantly reduce the ability of individuals of a population to hybridise with individuals of other populations.
- **crossing group(s)**
  any group of individuals that comprises a unique set of parents: (1) diallel crossing group—controlled crosses are made between each pair of parents in the group but crosses with parents outside the group are excluded, (2) factorial crossing group—a limited number of parents are used as male testers in controlled crosses with an unlimited number of female parents, (3) open-pollinated crossing group—all parents in a breeding population are included in a progeny test or series of tests.
- **crossing over**
  a process in which homologous chromosomes exchange parts normally reciprocally but sometimes unequally. The exchange of corresponding chromosome parts between homologues by breakage and reunion of DNA molecules normally during prophase I of meiosis but also occasionally during mitosis.

- **crossing-over map**
  a genetic map made by utilising crossing-over frequencies as a measure of the relative distances between genes in one linkage group (chromosome).

- **cross-inoculation**
  inoculation of one legume species by the symbiotic bacteria from another.

- **crossover suppression**
  reduction of crossing over within an inversion loop in inversion heterozygotes due to physical constraints during synapsis. Crossing over within an inversion loop, when it does occur, leads to defective (deleted and duplicated) crossover chromosomes and mortality of zygotes carrying them.

- **crossovers**
  the exchange of genetic material between two paired chromosome during meiosis.

- **cross-pollination**
  the transfer of pollen from the stamen of a flower to the stigma of a flower of a different genotype but usually of the same species, with subsequent growth of the pollen tube.

- **cross-protection**
  plant protection conferred on a host by infection with one strain of, for example, a virus that prevents infection by a closely related strain.

- **cross-resistance**
  resistance associated with a change in one genetic factor that results in resistance to different chemical pesticides that were never applied.

- **cross-sterility**
  the failure of fertilisation because of genetic or cytological conditions (incompatibility) in crosses between individuals crossing barrier.

- **crown**
  the stem-root junction of a plant (e.g., the overwintering base of an herbaceous plant) , the term is also used for the treetop.

- **crucifer**
  a plant belonging to the Brassicaceae or mustard family, a large dicotyledonous family of important crop and ornamental plants (turnip, cabbage, etc.)

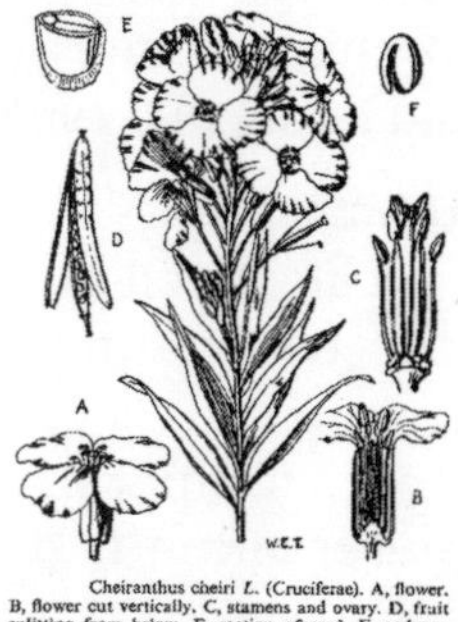

Cheiranthus cheiri *L.* (Cruciferae). A, flower. B, flower cut vertically. C, stamens and ovary. D, fruit splitting from below. E, section of seed. F, embryo. (After Baill.)

- **cruciform configuration**
a region of DNA having a sequence at one end repeated but inverted at the other end, so that each strand may pair with itself to form a helix extending sideways from the main helix.

- **crust**
a surface layer of soil that becomes harder than the underlying horizon when dry.

- **cryoability**
the ability of plant material (seeds, tissue, organs) to preserve or store under very low temperatures, usually in liquid nitrogen (–196°C).

- **cryobank**
the preservation or storage under very low temperatures, usually in liquid nitrogen (–196°C).

- **cryodamage**
damage caused by exposure to cold conditions.

- **cryopreservation**
the preservation or storage in very low temperatures, usually in liquid nitrogen (–196°C).

- **cryoprotectant**
a chemical, which is used to protect seeds, cultured material, tissue, organs or cells from the low temperature in cryopreservation (e.g., glycerol).

- **cryoscopy**
a technique for determining the molecular weight of a substance by dissolving it and measuring the freezing point of the solution.

- **cryostat**
a device designed to provide low-temperature environments in which experiments may be carried out under controlled conditions.

- **cryptic colouration**
colouration that allows an organism to match its background and hence become less vulnerable to predation or recognition by prey.

- **cryptochrome**
photomorphogenesis.

- **cryptogam(ous)**
reproduction by spores or gametes rather than seeds.

- **cryptomerism**
the phenomenon that a gene or an allele does not show a phenotypic effect unless it is activated by another genetic factor, which leads to a sudden change of qualities in the progeny not recognised among the ancestors.

- **crystallography**
the determination of, for example, the protein structure. The

protein is crystallised and the crystals examined using X rays. The diffraction angles of the X rays are used to compute the relative positions of components of the protein and thus its structure.

- **$CsCl_2$ gradient**
a method used to separate DNA or phages according to buoyant density.

- **cuckoo chromosome**
an alien chromosome that shows a preferential transmission during generative reproduction, found in certain wheat-*Aegilops* crossing progenies.

- **cuckoo gene**
it refers to a gene conferring preferential transmission from the maternal parent.

- **cucullate**
hood- or cowlike in form.

- **culled**
off-grade.

- **culling**
the postharvest removal of pathogen-infected or damaged fruit, seeds or plants by screening procedures. The culled or off-graded material can later be individually analysed or discarded.

- **culm**
the jointed stem in cereals, grasses or sedges, filled with pith or solid.

- **cultigen**
a cultivated plant or group of plants for which there is no known wild ancestor (e.g., maize *Zea mays*).

- **cultivar**
see **plant variety**.

- **cultivation**
the art or process of agriculture.

- **cultivator**
an implement drawn between rows of growing plants to loosen the earth and destroy weeds.

- **culture**
1. tissue or cells multiplying by asexual division, grown for experimentation.
2. A growth of one organism or of a group of organisms for the purpose of production, trade and utilisation or for experiments.

- **culture collection**
a collection of cultures of more or less defined or characterised viruses, bacteria and other organisms, usually used for reference and comparison with new isolates.

- **culture medium**
medium on or in which tissues, organs or cells are cultured. It supplies the mineral and hormonal requirements for the growth.

- **culture tube**
a tube in which tissue, organs, cells or organisms are cultured.

- **cumulative genes**
polymeric non-allelic genes.

- **cupule**
a cup-like structure at the base of some fruits.

- **cushion plants**
have small, hairy or thick leaves borne on short stems and form a tight hummock.

- **cut-and-come-again**
applied to any plant that is cut or sheared after flowering and blooms again (e.g., *Petunia* spp., pansy).

- **cuticle**
a thin, waxy, protective layer covering the surface of the leaves and stems.

- **cutin**
the complex mixture of fatty-acid derivatives with water-proofing qualities of which the cuticle is composed.

- **cutinise**
to impregnate a cell or a cell wall with cutin, a complex fatty or waxy substance, which makes the cell more or less impervious to air and moisture.

- **cutout**
the occurrence of physiologically indeterminate growth (e.g., in cotton).

- **cutting**
a section of a plant that is removed and used for propagation, cuttings may consist of a whole or part of a stem (leafy or nonleafy), leaf, bulb or root. A root cutting consists of root only, other cuttings have no roots at the time they are made and inserted. As opposed to division, a kind of propagation that consists of part of the crown of a plant or of its above-ground portion and roots.

- **CV.**
cultivar.

- **cybrid**
the hybrid formed from the fusion of a cytoplast and a whole cell. The cytoplast may transmit cytoplasmic components independently of the cell genome.

- **cyclic AMP (camp)**
a form of AMP (Adenosine Monophosphate) used frequently as a second messenger in eukaryotics and in catabolite repression in prokaryotes.

- **cyclical parthenogenesis**
a life history in which a sequence of apomictic genera-

tions is followed by amphimictic generations.

- **cycling**
  a round of recombination, testing and selection, it may often refer to the mating breed.

- **cycling strategy**
  choice of methods of recombination, testing and selection in repeated cycling breeding strategy.

- **cycloheximide (actidione)**
  an antibiotic from *Streptomyces griseus*, antibacterial and antifungal.

- **cyme**
  an inflorescence in which each axis ends in a flower.

- **cysteine (cys)**
  an aliphatic, polar alpha-amino acid that contains a sulphydryl group.

- **cystic fibrosis**
  an autosomal recessive genetic condition of the exocrine glands which causes the body to produce excessively thick, sticky mucus that clogs the lungs and pancreas interfering with breathing and digestion.

- **cystic fibrosis (cf)**
  a potentially lethal human disease of secretory cells, showing excess lung mucus secretion and inherited as an autosomal recessive on chromosome 7. CF is caused by mutations in a gene encoding the cystic fibrosis membrane conductance regulator, a transmembrane protein involved in ion transport.

- **cytidine**
  the nucleoside containing cytosine as its base.

- **cytochimera**
  different tissues or parts of them differ in chromosome number.

- **cytodifferentiation**
  the sum of processes by which, during the development of the individual zygote, specialised cells, tissue and organs are formed.

- **cytogamy**
  the fusion or conjugation of cells.

- **cytogenetic map**
  a map showing the locations of genes on a chromosome.

- **cytogenetics**
  the cytological approach to genetics, mainly involving microscopic studies of chromosomes.

- **cytogony**
  the reproduction by single cells.

- **cytohet**
  a cell containing two different cytoplasmic genomes (e.g., mi-

tochondria) that differ in one or more genes contributed by two parents. Thus, the individual is cytoplasmatically heterozygous.

- **cytokinesis**
during the division of a cell, the division of the constituents of the cytoplasm. It usually begins in early telophase with the formation of a cell plate, which is assembled within the phragmoplast across the equatorial plane. The phragmoplast is a complex array of GOLGI-derived vesicles, microtubule, microfilaments and endoplasmatic reticulum that assembles during the late anaphase and is dismantled upon completion at the new wall.

Cytokinesis

Animal Cell

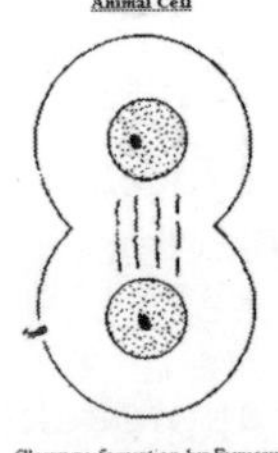

Cleavage formation by Furrow

Plant Cell

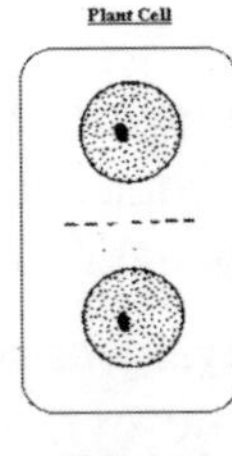

Cell place formation

- **cytokinin**
one of a group of hormones, including kinetin, that act synergistically with auxins to promote cell division but, unlike auxins that promote lateral growth.

- **cytology**
the branch of biology dealing with the structure, function and life history of the cell.

- **cytolysis**
breaking up or solution of the cell wall.

- **cytomixis**
the extrusion or passage of chromatin from one cell into the cytoplasm of an adjoining cell.

- **cytoplasm**
the material between the nuclear and cell membranes, includes fluid (cytosol) organelles and various membranes, but excluding the nucleus.

- **cytoplasmic inheritance**
1. inheritance via genes found in cytoplasmic organelles. Extra-chromosomal inheritance controlled by non-nuclear genomes.
2. A non-Mendelian (extra-chromosomal) inheritance via genes in cytoplasmic organelles (mitochondria, plastids).

- **cytoplasmon**
all cytoplasmic hereditary constituents of a cell excepting those localised in the plastids and mitochondria.

- **cytoplasm-restorer**
cytoplasmic male sterility.

- **cytoplast**
  the cytoplasm as a unit, as opposed to the nucleus.

- **cytosine (c)**
  a pyrimidine base that occurs in both DNA and RNA.

- **cytosol**
  the fluid portion of the cytoplasm, outside the organelles.

- **cytostatic**
  any physical or chemical agent capable of inhibiting cell growth and cell division.

- **cytotaxonomy**
  the study of natural relationships of organisms by a combination of cytology and taxonomy.

- **cytotoxic t lymphocyte (TC)**
  lymphocyte responsible for attacking cancerous host cells or cells infected with an invading bacterium or virus.

- **cytotype**
  any variety of a species whose chromosome complement differs quantitatively or qualitatively from the standard complement of that species.

- **dam gene**
  DNA adenine methylation gene of *Escherichia coli* it methylates the sequence GATC (*Sau3A* cleaves methylated and unmethylated DNA, *MboI* cleaves only unmethylated DNA, *DpnI* cleaves only the methylated sequence).

- **dammar resin**
  a hard, lustrous resin derived from Asian trees of the monkey-puzzle family.

- **dandelion weeder**
  asparagus knife.

- **dark field microscopy**
  a microscope designed so that the entering centre light rays are blacked out and the peripherical rays are directed against the object from the side. As a result the object being viewed appears bright upon a dark background.

- **dark reaction**
  the phase of photosynthesis, not requiring light, in which carbohydrates are synthesised from carbon dioxide.

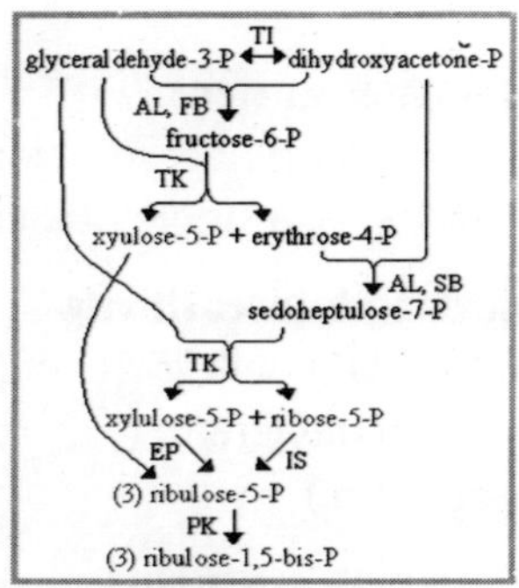

- **dark respiration**
  dark reaction.

- **Darwinian fitness**
  the relative probability of survival and reproduction for a genotype. see **fitness**.

- **Darwinism**
  the theory that the mechanism of biological evolution involves natural selection of adaptive variations.

- **dauermodification**
  the persistence for several generations of an environmentally induced trait.

- **daughter cell**
  the cells resulting from the division of a single cell.

- **daughter chromosome**
  any of the two chromatids of which the replicated chromosome consists after mitotic metaphase or anaphase II of meiosis.

- **daughter nucleus**
  the nuclei that result from the division of a single nucleus.

- **daylength insensitivity**
  these plants will flower independent of daylength (e.g., tomato or cotton.)

- **daylength sensitivity**
  plants that will flower only when the daily photoperiod shows a critical length.

- **day-neutral plants**
  no daylength requirement for floral initiation.

- **debearder**
  in seed precleaning procedures, it has a hammering or flailing action that removes awns, beards or lint from seed and tends to break up seed clusters of the chaffy grasses, as well as multiple seed units of nonchaffy forms.

- **decarboxylase**
  an enzyme that removes the carboxyl group from an organic compound.

- **decay**
  the destruction of plant material by fungi and bacteria.

- **decentralised plant breeding (programme)**
  a well-defined set of breeding experiments carried out in a variety of local sites (communities, farmer fields) that represent real farming conditions, as opposed to a single, central, research station site that does not represent a real farming context.

- **deciduous**
  a plant whose leaves are shed at a season or growth stage.

- **decimal code for the growth of cereal plants**
  a decimal code used for describing different growth stages

of cereal plants, it is applied for comparison of several morphological stages. There are several schemes of description.

- **declining vitality**
seeds that are aged or have been subjected to unfavourable storage conditions. They usually show a slow germination. Some of the essential plant parts are frequently stunted or lacking. Saprophytic fungi may also interfere with the growth of the seedlings.

- **decompose**
to rot or putrefy.

- **decomposer**
an organism that breaks down dead tissues into simple chemical components, thereby returning nutrients to the physical environment of plants.

- **decondensation stage**
a stage between interphase and prophase of mitosis in which heterochromatin is decondensed for a short period.

- **decontaminate**
to free from contamination, purify.

- **decumbent**
lying on the ground with the end ascending.

- **decurrent**
describes the open type of collar in barley where the margin of the platform is incomplete, merging with the neck, in general, extending downward from the point of insertion.

- **dedifferentiation**
a loss of specialisation of a cell. It can be observed when differentiated cells are placed in vitro culture.

- **defence response**
the active response of a plant to pathogen attack, it includes the elements that inhibit pathogen development. A defence response is activated in compatible and incompatible interactions.

- **deficiency**
the absence of part of the normal genome or chromosome set. See **deletion**.

- **deficiency disease**
any disease caused by the lack or insufficiency of some nutrient, element or compound (e.g., copper. Zinc or iron deficiency in cereals.)

- **definite host**
the host in which the parasite attains sexual maturity.

- **deflexed**
bent sharply downward.

- **defoil**
  to strip a plant of leaves.

- **defoliant**
  a chemical or method of treatment that causes only the leaves of a plant to fall off or abscise (e.g., it is applied before harvest of potato.)

- **defoliation**
  the process of leaves being removed from a plant (e.g., to make harvest easier).

- **degeneracy of genetic code**
  degenerated code.

- **degenerate code**
  a code in which several code words have the same meaning. The genetic code is degenerate because there are many instances in which different codons specify the same amino acid. A genetic code in which some amino acids may each be encoded by more than one codon.

- **degermed**
  grains from which the embryo (germ) has been removed.

- **degradation**
  the progressive decrease in vigour of successive generations of plants, usually caused by unfavourable growing conditions or diseases. Viruses may cause great loss of vigour. In agriculture, the change of one kind of soil to a more highly leached soil.

- **degree of dominance**
  dominance.

- **degree of freedom (DF)**
  the number of items of data that are free to vary independently, in a set of quantitative data. For a specified value of the mean, only (n −1) items are free to vary, since the value of the nth item is then determined by the values assumed by the others and by the mean.

- **degree of genetic determination**
  the portion of total variance that is genetically determined.

- **dehiscence**
  the bursting open at maturity of a pod or capsule along a definite line or lines.

- **dehiscent fruit**
  dehiscence.

- **dehull**
  removal of outer seed coat (hull) or to remove the glumes of cereal caryopses.

- **dehusk**
  to remove the husk (e.g., the leaf sheath of a maize cob or the outer layer of a coconut).

- **dehydration**
the elimination of water from any substance.

- **dehydrogenase**
an enzyme that catalyses the removal of hydrogen from a substrate.

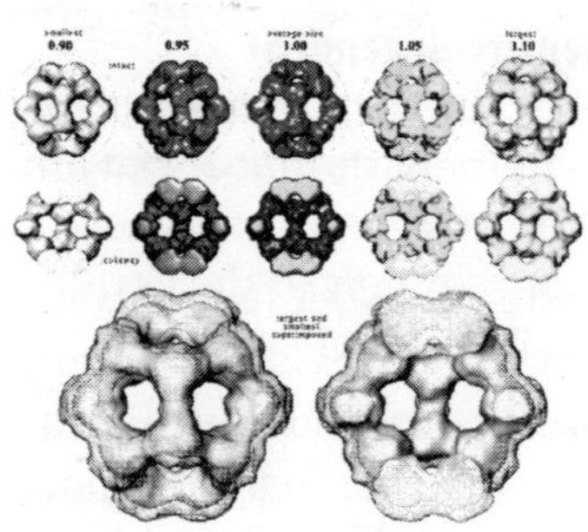

- **deletion**
loss of a DNA (chromosome) segment from a chromosome. Deletions are recognised genetically by:
1. absence of reverse mutation.
2. Presence of a deletion loop at meiosis visualised cytologically.
3. revealation of recessive lethals.
4. pseudodominance.

- **deletion chromosome**
a chromosome containing a deletion.

- **deletion mapping**
the use of overlapping deletions to localise the position of an unknown gene on a chromosome or linkage map.

- **deletion mutation**
a mutation in which one or more bases are removed from the DNA sequence of a gene.

- **deltoid**
shaped like the Greek letter 'delta'.

- **deme**
a locally interbreeding population.

- **demic selection**
special type of intergroup selection that does not necessarily involve direct competition. It has an effect on the general genetic composition of a population if subsets of a population have different gene frequencies.

- **denaturation**
reversible or irreversible alterations in the biological activity of proteins or nucleic acids that are brought about by changes in structure other than the breaking of the primary bonds between amino acids or nucleotides in the chain.

- **denaturation map**
a map of a stretch of DNA showing the locations of local denaturation loops, which correspond to regions of high AT content.

- **denatured**
  loss of natural configuration (of a molecule) through heat or other treatment. Fully denatured DNA is single-stranded.

- **denatured protein**
  a protein whose properties have been altered by treatment with physical or chemical agents.

- **dendrogram**
  a genealogical diagram that resembles a tree. An evolutionary tree diagram may order objects, individuals genes, etc., on the basis of similarity cluster analysis.

- **dendrology**
  the branch of botany dealing with trees and shrubs.

- **denitrification**
  the conversion of nitrate or nitrite to gaseous products, chiefly nitrogen and/or nitrous oxide, resulting in the loss of nitrogen into the atmosphere and therefore undesirable in agriculture.

- **density gradient centrifugation**
  the separation of macromolecules or subcellular particles by sedimentation through a gradient of increasing density under the influence of a centrifugal force. The density gradient may either be formed before the centrifugation run by mixing two solutions of different density (e.g., in sucrose density gradients) or it can be formed by the process of centrifugation itself (e.g., in $CsCl_2$ and Cs_2SO_4 density gradients).

- **density-dependent**
  of or referring to a factor, such as nutrient shortage, that limits the growth of a population more strongly as the density of the population increases.

- **density-gradient centrifugation**
  a method of separating macromolecules by their
  (1) differential rate of sedimentation in a centrifugal gradient
  (2) differential bouyancy in a density gradient.

- **density-independent**
  of or referring to a factor, such as weather, that can limit the size of a population but does not act more strongly as the density of the population increases.

- **dent corn**
  a variety of maize, *Zea mays* ssp. *indentata*, having yellow or white kernels that become indented as they ripen.

- **deoxyribonuclease**
  any of several enzymes that break down the DNA molecule into its component nucleotides.

- **deoxyribonucleic acid (DNA)**
  a nucleic acid, characterised by the presence of a sugar deoxyribose, the pyrimidine bases cytosine and thymine and the purine bases adenine and guanine. Its sequence of paired bases constitute the genetic code.

- **deoxyribonucleotide**
  see **nucleotide.**

- **depauperate fauna**
  a fauna, especially common on islands, lacking many species found in similar habitats elsewhere.

- **dephosphorylation**
  the removal of a phosphate group from an organic compound, as in the changing of ATP to ADP.

- **deployment**
  the physical movement of clones (ramets) or other genetic units from one site (usually a nursery) to another (usually plantations), often including their spatial configuration on the recipient site.

- **derepressed**
  the condition of an operon that is transcribing because repressor control has been lifted. May apply more generally to any gene being transcribed.

- **derivative hybrid**
  a hybrid arising from a certain cross between two hybrids.

- **dermal tissue system**
  the tissue system in plants that forms their outer covering.

- **dermatogen**
  a specialised meristem in flowering plants in which floral induction begins, gives rise also to the epidermis.

- **descendant**
  an individual resulting from the sexual reproduction of one parental pair of individuals.

- **descent**
  the act, process or fact of descending.

- **descriptor**
  an identifiable and measurable characteristic used to facilitate data classification, storage, retrieval and use.

- **desiccant**
  a chemical applied to crops that prematurely kills their vegetative growth. Often used for legume seed crops so the seed can be harvested prior to normal plant senescence.

- **desiccation**
  the process of drying out.

■ **desiccation-tolerant seed**
there are basically two types of seed: (a) desiccation-tolerant and (b) desiccation-intolerant. Most of the plants produce desiccation-tolerant seeds, which means they can be safely dried for long-term storage, exceptions include many aquatic plants, large-seeded plants and some trees (oaks, buckeyes etc.), any of which produce desiccation-intolerant seeds which will die if allowed to dry. They do not enter dormancy after maturing, instead, respiration and other physiological processes continue.

■ **desinfectant**
a physical or chemical substance for the destruction of pathogenic micro-organisms.

■ **desmosome**
a plaquelike site on cell surfaces that function in maintaining cohesion with an adjacent cell.

■ **desynapsis**
the premature separation of paired chromosomes during diplotene or diakinesis of meiotic prophase. It is often genetically controlled but can also be induced by special environmental conditions (e.g., heat).

■ **detasseling**
artificially removing (cutting or pulling) the tassel of the female parent to prevent seifing during hybrid seed maize production.

■ **detergent**
any synthetic organic cleaning agent that is liquid or water-soluble and has wetting-agent and emulsifying properties.

■ **determinant**
a spatially localised molecule that causes cells to adopt a particular fate or set of related fates.

■ **determinate**
descriptive of an inflorescence in which the terminal flower opens first, thus arresting the prolongation of the floral axis.

■ **determination**
the process of commitment of cells to particular fates.

■ **deterministic**
referring to events that have no random or probabilistic aspects but proceed in a fixed predictable fashion.

■ **deterministic process**
stochastic.

■ **detoxication**
the metabolic process by which toxins are changed into less toxic or more readily excreted substances.

- **detritivore**
  a consumer that relies on dead tissues for nutrients.

- **development**
  the process whereby a single cell becomes a differentiated organism. The process of orderly change that an individual goes through in the formation of structure.

- **developmental cycle**
  the gradual progression of phenotypic modifications of an organism during development.

- **developmental genetics**
  the study of mutations that produce developmental abnormalities in order to gain understanding of how normal genes control growth, form, behaviour, etc.

- **developmental stage**
  growth stages.

- **devernalisation**
  the reversion of vernalisation by nonvernalising temperatures or other means.

- **deviation**
  the departure of a quantity from its expected value.

- **dextrose**
  an aldohexose monosaccharide that is a major intermediate compound in cellular metabolism. The dextrorotatory form of glucose, occurring in fruits and commercially obtainable from starch by acid hydrolysis.

- **DH lines**
  doubled-haploid lines.

- **diadelphous**
  showing stamens united in two sets by their filaments (e.g., in pea and bean flowers), nine out of ten stamens are usually united while one is by itself.

- **diakinesis**
  the final stage of prophase 1 of meiosis when chiasmata terminalise.

- **diallel**
  in either the complete or incomplete diallel, identities of both seed and pollen parents are maintained for each family.

- **diallel cross**
  the crossing in all possible combinations of a series of genotypes.

- **diallel crossing group**
  crossing group(s).

- **diallel mating**
  diallel cross.

- **diapause**
  a period of dormancy , in insects, a state during which growth and development is temporarily arrested.

- **dibble**
a pointed tool used to make holes in the ground for seeds and seedlings pricking-out peg.

- **dibble planting**
dibbling.

- **dibbling (seed)**
sowing in holes made by a pointed tool.

- **dicentric chromosome**
a chromosome with two centromeres.

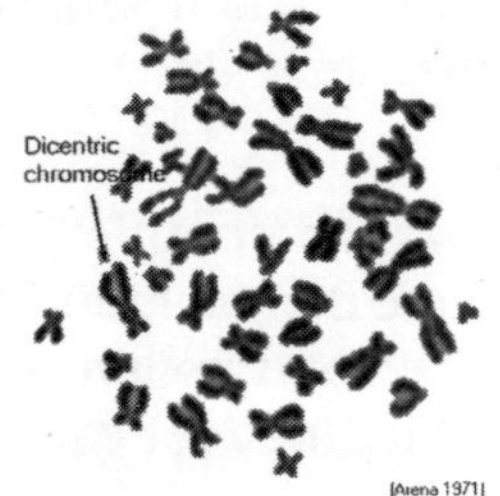

(Arena 1971)

- **dichogamy**
the condition in which male and female parts of a flower mature at different times.

- **dichophase**
the phase of the mitotic cycle in which a cell is determined for further mitotic differentiation or for special cell functions.

- **dichotomous ramification**
branching, frequently successively, into two more or less equal arms.

- **diclinous species**
species having pistils and stamens on different flowers.

- **dicot**
an abbreviated name for dicotyledon, which refers to plants having two seed leaves.

- **dicotyledons**
plant species having two cotyledons, flower parts arranged in fours or fives or multiples thereof, net-veined leaves and vascular bundles in the stem arranged in a ring.

- **didiploid**
two different diploid chromosome sets present in one cell or organism.

- **differential centrifugation**
a method(s) of separating subcellular particles by centrifugation of cell extracts at successively higher speeds. It is based on differences in sedimentation coefficients that are roughly proportional to particle size, in other words, large particles (nuclei, chloroplasts or mitochondria) are sedimented at lower speeds than small particles (ribosomes).

- **differential selection**
the difference between a selected plant, family or clone and the average of the population from which it is taken.

- **differential staining**
  in microbiology, staining procedures that divide bacteria into separate groups based on staining properties. in cytology, staining procedures that divide chromosomes or genomes into separate segments based on structural and biochemical properties.

- **differential variety**
  a host variety, part of a set differing in disease reaction, used to identify physiologically specialised forms of pathogen. see **differential host.**

- **differentiation**
  the changes in cell shape and physiology associated with the production of the final cell types of a particular organ or tissue.

- **differentiation of cells**
  the development of specialised kinds of cells from nonspecialised cells in a growing tissue.

- **diffuse stage**
  a meiotic prophase stage in which the chromosomes become reorganised, they may almost disappear.

- **diffusion**
  passive movement of molecules from areas of high concentration of the molecule to areas of low concentration.

- **dig**
  to break up and turn over piecemeal the soil or ground.

- **digametic**
  heterogam(et)ic.

- **digenic**
  it refers to an inheritance that is determined by two genes.

- **digestibility**
  the attribute of forage biomass to be digested by grazing animals—an important selection criterion in forage crop breeding.

- **digitate**
  fingerlike or a compound, with the members arising together at the apex of the support.

- **digoxigenin (dig)**
  antigenic alkaloid from *Digitalis* spp. (foxglove), which is used to label DNA in situ hybridisation.

- **dihaploid**
  a haploid cell or individual containing two haploid chromosome sets—not to be confused with doubled-haploid.

- **dikaryon**
  a dinucleate cell.

- **dimer**
a protein that is made up of two polypeptide chains or subunits paired together.

- **dimerisation**
the chemical union of two identical molecules.

- **dimethyl sulfoxide (DMSO)**
a liquid solvent, C_2H_6OS, approved for better penetration of specific substances through the cell wall.

- **dimorph**
dimorphism.

- **dimorphism**
the occurrence of two forms of individuals within one population or other taxa (e.g., sexual dimorphism or the presence of one or more morphological differences that divide a species into two groups.)

- **dioecious**
possessing male and female flowers or other reproductive organs on separate, unisexual, individual plants (e.g., in hemp or spinach.)

- **dioecious plant**
a plant species in which male and female organs appear on separate individuals.

- **dioecism**
the phenomenon of plants showing either male or female sex organs.

- **diphasic**
chromosomes that show both euchromatic and heterochromatic segments.

- **diplandroid**
in parthenogenesis, the progeny derives from an unreduced sperm cell.

- **diploid**
the state of having each chromosome in two copies per nucleus or cell. A cell having two chromosome sets or an individual having two chromosome sets in each of its cells.

- **diploidisation**
in polyploids, a natural or induced mechanism in which the chromosomes pair completely or partially as bivalents, although polyploid sets of chromosomes are present. It may be caused by a structural differentiation of homologous chromosome sets or by genetic control, for example, in bread wheat three homoeologous genomes are available (AABBDD), they do not pair as hexavalents but exclusively as bivalents.

- **diplonema (diplotene stage)**
the stage of prophase of meiosis I in which chromatids appear to repel each other.

- **diplontic**
diploid.

- **diplophase**
the diploid generation phase after fertilisation to meiosis.

- **diplospory**
a type of agamospermy (apomixis) in which a diploid embryo sac is formed from archesporial origin.

- **diplotene**
the stage in the prophase of first meiosis, when the paired homologous chromosomes separate except where they are held together by chiasmata.

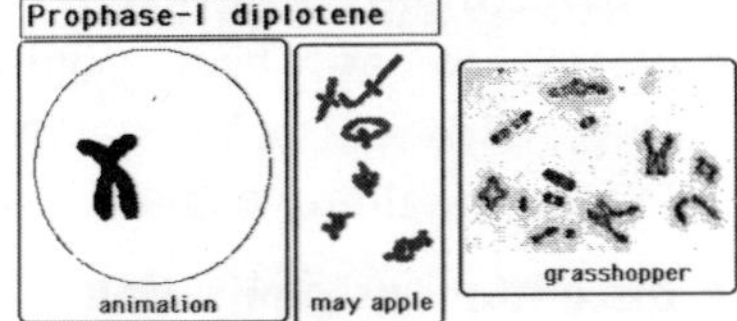

Chromosomes are are shorter

Homologous chromosomes are **repelling** each other

Chiasmata appear

- **dipping**
the immersion of seedling roots in a solution or water prior to planting.

- **direct embryogensis**
embryoid formation directly on the surface of zygotic or somatic embryos or on seedling plant tissues in culture without an intervening callus phase.

- **direct organogenesis**
organ formation directly on the surface of relatively large intact explants without an intervening callus phase.

- **directed dominance**
see **directional dominance dominance**.

- **directed mutagenesis**
altering some specific part of a cloned gene and reintroducing the modified gene back into the organism.

- **directional cloning**
in biotechnology, DNA inserts and vector molecules are digested with two different restriction enzymes to create noncomplementary sticky ends at either end of each restriction fragment. It allows the insert to be ligated to the vector in a specific orientation and prevents the vector from recircularising.

- **directional dominance**
a type of dominance in which the majority of dominant alleles have positive effects in one direction.

- **directional mutation**
a genetic change that favours a certain genotype or population.

- **directional selection**
a type of selection that removes individuals from one end of a

phenotypic distribution and thus causes a shift in the distribution. Selection that changes the frequency of an allele in a constant direction, either toward or away from fixation for that allele.

- **dirty seed**
  endophyte-infected seeds (e.g., in grass).

- **disaccharide**
  any of a group of carbohydrates, such as sucrose, that yield monosaccharides on hydrolysis.

- **disassortative mating**
  the mating of two individuals with dissimilar phenotypes.

- **disbud**
  removing buds, shoots or growing tips (with finger and thumb) of a tree, vine or flowering plant to encourage production of sideshoots or high-quality flowers and fruits, pinching out is also used when small side shoots are completely removed. It is done when single stems are desired, especially when training to form the 'trunks' of standard (tree-form) specimens.

- **disc floret**
  a small flower, usually one of a dense cluster (e.g., sunflower).

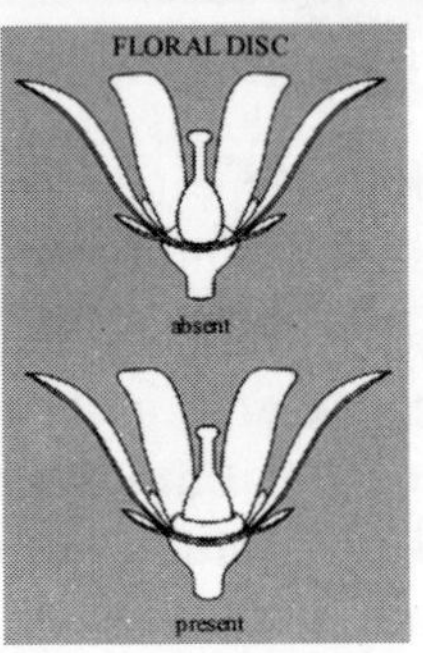

- **disc grain-grader**
  discs revolve through a seed mass and a certain size of seeds are lifted and discharged, while the other size (e.g., the longer ones) is rejected by the disc indents.

- **discontinuous character**
  variation in which discrete classes can easily be recognised (e.g., flower colour, straw length.) qualitative character.

- **discontinuous replication**
  replication of DNA in short 5' to 3' segments using the 5' to 3' strand as a template while going backward away from the replication fork.

- **discontinuous variation**
  variation that falls into discrete categories (e.g., the colour of garden peas.)

- **discrete generations**
  generations that have no overlapping reproduction. All reproduction takes place be-

tween individuals in the same generation.

- **disease**
a condition in which the use or structure of any part of a living organism is not normal, harmful deviation from normal functioning of physiological processes. Six types of causal agents can be considered: (1) fungi, (2) bacteria, (3) viruses, (4) nematodes, (5) insects and (6) plant parasites.

- **disease avoidance**
avoiding the disease by, for example, growing the crop sufficiently early so that the vulnerable part of the plant's growing cycle is over before the disease-causing organism arrives in the area, this stops the disease from starting. It is sometimes called 'passive resistance'.

- **disease control**
several types of disease control can be classified: (1) disease resistance, (2) protection, (3) avoidance, (4) exclusion, (5) eradication, (6) therapy.

- **disease cycle**
a cyclical sequence of host and parasite development and interaction that result in disease and reproduction of the pathogen.

- **disease eradication**
this control measure is applied to a situation in which the disease is present in the area. It involves removing the disease by, for example, burning all stubble of the diseased crop, in order to prevent transfer of the disease from the previous crop to the next season.

- **disease protection**
involves protecting the plant with a chemical. This is applied before the disease starts and prevents the beginning of problems. Systemic fungicides can penetrate and move inside the plant, they therefore have a greater exposure to the pathogenic organism, some are taken up by roots, other by leaves.

- **disease resistance**
the ability to resist disease or the agent of disease and to remain healthy.

- **disease therapy**
this applies to removing the particular part of the plant that is diseased, often applied to large and valuable plants, such as fruit trees, where a diseased branch may be removed by a tree surgeon.

- **disease triangle**
the three conditions required for a disease to occur, namely: a susceptible host, a sui pathogen and an appropriate environment.

- **disinfectant**
  a chemical treatment used to disinfect seed for planting, it is commonly useful for surface-borne pathogens.

- **disjunction (of daughter chromosomes)**
  the separation of homologous chromosomes at the anaphase stage of mitosis and meiosis and movement toward the poles of nuclear spindle.

- **disjunctional separation**
  alternative disjunction.

- **disk**
  enlarged growth of the head made up of a circular arrangement of fused petals (e.g., in sunflower).

- **disk plough**
  a plough with saucer-shaped units for breaking the soil.

- **dislocation**
  the displacement of a chromosome segment away from its original position in the chromosome.

- **dispermy**
  the entering of two sperm cells into one egg cell.

- **dispersal**
  the spread of a pathogen within an area of its graphical range.

- **dispersing agent**
  a chemical added to a pesticide formulation to aid the efficient distribution of particles of the active ingredient.

- **dispersive replication**
  disproved model of DNA synthesis suggesting more-or-less random interspersion of parental and new segments in daughter DNA molecules. A postulated mode of DNA replication combining aspects of conservative and semiconservative replication, known to be incorrect.

- **disporic**
  having two spores.

- **disruptive selection**
  a type of selection that removes individuals from the centre of a phenotypic distribution and thus causes the distribution to become bimodal.

- **dissecting microscope**
  usually, a low-power microscope (50x magnification) used to facilitate dissection, examination or excision of small plant parts. However, in recent biotechnology high-power microscopes also are applied for dissections.

- **dissemination**
  the spread of seeds or spores dispersal.

- **dissepiment**
  a partition within an organ of a plant (e.g., the membrane that separates sections of the orange and other citrus fruits.)

- **dissimilation**
  assimilation.

- **distal**
  farthest from the point by which it is attached to the starting point.

- **distance isolation**
  several plants can be protected from cross-pollination by separation, by a certain distance, such as lettuces (~25 feet) or eggplants (~50 feet), insect-pollinated plants, such as the cabbage family (collards, broccoli, etc.), squashes and okra require from ¼ to 1 mile for complete safety, maize, a wind-pollinated plant, can require a mile or more for safe distance isolation and members of the beet family may need as many as 5 miles, the exact distance for safely isolating a particular crop depends on a number of factors, e.g., the type of plant and how it is pollinated (i.e., wind, insects, selfing or a combination of these), the location, climate, prevailing wind patterns and surrounding terrain and vegetation features, the relative size of plantings etc.

- **distant hybridisation**
  the crossing and/or hybridisation of members of different genera.

- **distichous**
  in two vertical ranks.

- **distinguishable hybrid**
  a type of hybrid in which intermediate inheritance is phenotypically expressed (i.e., the heterozygous gene constitution is visible by the phenotype.)

- **distribution**
  see **statistical distribution**.

- **distribution function**
  a graph of some precise quantitative measure of a character against its frequency of occurrence.

- **distyly**
  the presence of either pin or thrum flowers, pin x pin and thrum x thrum crosses are incompatible due to alleles at a single locus, the thrum morphology is controlled by a dominant allele *S* and the pin morphology by the recessive alleles.

- **disulphide bridge**
  a covalent bond formed between two sulphur atoms, it is a particular feature of peptides and proteins, where it is formed between the sulphydryl groups of two cysteine residues, help-

ing to stabilise the tertiary structure of these compounds.

- **ditelocentric**
  ditelosomic.

- **ditelomonotelosomic**
  a cell or an individual that has a pair of telocentric chromosomes for one arm and a single telocentric chromosome from the other arm.

- **ditelosomic**
  a cell or an individual that has two telocentrics of one chromosome arm.

- **ditelotrisomic**
  a cell or an individual that has two telocentric chromosomes of one arm plus one complete chromosome of the homologue See **aneuploid**.

- **ditertiary compensating trisomic**
  a cell or an individual with a compensating trisomic chromosome in which a missing chromosome is compensated by two tertiary chromosomes.

- **diurnal rhythm**
  circadian rhythm.

- **divalent**
  bivalent.

- **divergent**
  set at an angle to one another herring-bone fashion, as in the lateral spikelets of the two-row barley spike.

- **diversifying selection**
  selection in which two or more genotypes show optimal adaptation under different environments.

- **division**
  a method of propagation by which a plant clump is lifted and divided into separate pieces, it includes roots and a growing point.

- **D-loop**
  configuration found during DNA replication of chloroplast and mitochondrial chromosomes wherein the origin of replication is different on the two strands. The first structure formed is a displacement loop or D-loop.

- **DMD**
  Duchenne Muscular Dystrophy.

- **DNA clone**
  a section of DNA that has been inserted into a vector molecule, such as a plasmid or a phage chromosome and then replicated to form many identical copies. See also **clone**.

- **DNA content**
  usually, the total DNA amount per nucleus, given as picograms.

- **DNA glycosylase**
  endonuclease initiating excision repair at various damaged or improper bases in DNA.

- **DNA gyrase**
  a topoisomerase that relieves supercoiling in DNA by creating a transient break in the double helix.

- **DNA hybridisation**
  base pairing of DNA from two different sources, in biotechnology, a technique for selectively binding specific segments of single-stranded (ss) DNA or RNA by base pairing to complementary sequences of ssDNA molecules that are trapped on a nitrocellulose membrane.

- **DNA library**
  clone library.

- **DNA ligase**
  an enzyme that closes nicks or discontinuities in one strand of double-stranded DNA by creating an ester bond between adjacent 3' OH and 5' PO4 ends on the same strand.

- **DNA methylation**
  the methylation of DNA bases by endogenic methylases.

- **DNA packing**
  the highly organised way in which large amounts of DNA are packed into the cells of eukaryotic organisms.

- **DNA polymerase**
  an enzyme that can synthesise new DNA strands using a DNA template. Several such enzymes exist. One of several classes of enzymes that polymerise DNA nucleotides using single or double-stranded DNA as a template.

- **DNA polymorphism**
  one of two or more alternate forms (alleles) of a chromosomal locus that differ in nucleotide sequence or have variable numbers of repeated nucleotide units.

- **DNA probe**
  a more or less defined piece of DNA that is used for DNA-DNA or DNA-RNA hybridisation experiments.

- **DNA repair**
  the reconstruction of DNA molecule after different sorts of DNA-strand damages by endogenic enzymes, it is involved in the recombination process and promotes the survival of an organism after partial DNA damage.

- **DNA replication**
  the process whereby a copy of a DNA molecule is made and thus the genetic information it contains is duplicated. The parental

double-stranded DNA molecule is replicated semiconservatively (i.e., each copy contains one of the original strands paired with a newly synthesised strand that is complementary in terms of AT and GC base pairing.)

- **DNA sequence**
  the relative order of base pairs, whether in a fragment of DNA, a chromosome or an entire genome. See **base sequence analysis**.

- **DNA sequencing**
  methods and procedures for determining the nucleotide sequence of a DNA fragment and/or chromosome.

- **DNA-amplified fingerprinting**
  a technology based on amplification of random genomic DNA sequences achieved by a single short (5-8 bases) oligonucleotide primer of arbitrary sequence. It produces a characteristic spectrum of short DNA pieces of varying complexity that are resolved on polyacrylamide gel (PAGE) following silver staining. It is used to detect genetic differences between genotypes as well as for detecting polymorphism even between organisms that are closely related, such as near isogenic lines. See **DNA fingerprint (ing)**.

- **DNA-DNA hybridisation**
  when DNA is heated to denaturation temperatures to form single strands and then cooled double helices will re-form (renaturation) at regions of sequence complementarity. This technique is useful for determining sequence similarity among DNAs of different origin and the amount of sequence repetition within one DNA.

- **DNASE**
  nuclease specific for DNA deoxyribonuclease.

- **domain**
  a discrete portion of a protein with its own function. The combination of domains in a single protein determines its overall function.

- **dome**
  in cereals, the zone of cells at the tip of the apical meristem (shoot apex), which, by cell division, forms the site for production of leaf and spikelet.

- **dominance**
  the quality of one of a pair of alleles that completely suppresses the expression of the other member of the pair when both are present, the degree of dominance is expressed by the ratio of additive genetic variance to total phenotypic vari-

ance. In case the ratio equals 1, the trait shows complete dominance, if the ratio is greater than 1, the trait shows overdominance, if the ratio is less than 1, the trait shows incomplete dominance.

■ **dominance hypothesis**
in hybrid breeding, dominant alleles of genes that should have stimulating effects on heterosis, while recessive ones should show inhibitory effects.

■ **dominance variance**
genetic variance at a single locus attributable to dominance of one allele over another.

■ **dominant**
an allele that determines phenotype even when heterozygous. Also the trait controlled by that allele.

■ **dominant epistasis**
one dominant factor A is epistatic of another factor B or B is hypostatic to A.

■ **dominant phenotype**
the phenotype of a genotype containing the dominant allele, the parental phenotype that is expressed in a heterozygote.

■ **donor plant**
the source plant used for propagation, crossing, etc., whether a simple individual, an explant, graft or cutting.

■ **dormancy**
a resting condition with reduced metabolic rate found in nongerminating seeds and nongrowing buds.

■ **dormant bud**
dormancy.

■ **dormant seeding**
sowing during late autumn or early winter after temperatures become too low for seed germination to occur until the following spring.

■ **dormant spray**
a pesticide applied to dormant, leafless plants to control insects and diseases.

■ **dorsal**
in general, upon or relating to the back or outer surface of an organ (abaxial), the side of the caryopsis on which the embryo is situated.

■ **dosage compensation**
a genetic process that compensates for genes that exist in two doses in the homozygous dominants, so that the heterozygotes produces the same amount of gene product as the homozygotes.

- **dosage effect**

  the influence upon a phenotype of the number of times a genetic element is present.

- **dose**

  see **gene dose**.

- **dose effect**

  dosage effect.

- **dot-blot analysis**

  a variant of the Southern transfer, different concentrations of RNA or DNA may be determined, nonradioactive DNA will be denatured and with different concentrations transferred to nitrocellulose filters. Only small dots are transferred, so that DNA may hybridise with radioactively labelled probe, after autoradiography. The intensity of blackness is used as a simple measure of DNA concentration or DNA homology.

- **double cropping**

  the more or less contemporary growing of two crops on the same field, for example, it might be to harvest a wheat crop by early summer and then plant maize or soybeans on that acreage for harvest in autumn. This practice is only possible in regions with long growing seasons.

- **double cross**

  a cross between two F1 hybrids, the method used for producing hybrid seed, four different lines (A, B, C, D) are used, A x B ¥ AB hybrid and C x D ¥ CD hybrid, the single-cross hybrids (AB and CD) are then crossed and the double-cross hybrid (ABCD) seed is used for the commercial crop.

- **double crossing-over**

  the situation in which two crossing-overs take place within a tetrad.

- **double digest**

  the product formed when two different restriction endonucleases act on the same sample of DNA.

- **double ditelocentric**

  the phenomenon in which both arms of a certain chromosome are present as telocentrics and each telocentric with its homologue.

- **double flower**

  flowers that have more than one row of petals, stamens and sometimes the pistils are transformed into petals or sometimes the petals split to form several more. Completely double flowers have usually lost their reproductive organs and are therefore unable to produce seeds. These varieties of plants are bred to stay fresh longer than single flowers when cut. Commonly, plants do not show double flowers, they can only be formed when three specific genes are simultaneously mutating. These genes are the main regulators for flower formation, their DNA sequences are almost identical. By induced mutations and subsequent combination of these genes double flowers can be induced (e.g., in *Arabidopsis*).

- **double helix**
the normal structural configuration of DNA consisting of two helices winding about the same axis. The structure of DNA first proposed by Watson and Crick with two interlocking helices joined by hydrogen bonds between paired bases.

- **double hybrid**
double cross.

- **double infection**
infection of a bacterium with two genetically different phages.

- **double monoisosomic**
the presence of two isochromosomes, one for each arm. It can derive from a double monotelosomic individual.

- **double monotelosomic**
the presence of two telocentrics, one for each arm.

- **double recessive**
an individual that is homozygous for a recessive allele (e.g., *aa*), as opposed to homozygous for a dominant allele (e.g., *AA*).

- **double reduction**
the condition in polyploids in which a heterozygous individual produces homozygous gametes.

- **double telotrisomic**
the presence of two telocentrics, one for each arm of a missing whole chromosome, but together with the complete homologue aneuploid

- **double-cross hybrids**
hybrids resulting from crossing two single cross hybrids.

- **doubled haploid**
a diploid plant, which results from spontaneous or induced chromosome doubling of a haploid cell or plant, usually after anther or microspore culture by using different means.

- **doubled haploid (DH) method**
  a method used to speed up the production of homozygotes and to decrease the population size for selection, in other words, generating haploid plants by parthenogenesis or by anther culture followed by doubling of the number of their chromosomes (spontaneously or induced) doubled-haploid lines (DH lines).

- **doubled-haploid lines (DH lines)**
  homozygous lines derived from haploidisation and doubling again the chromosome number doubled haploid (DH) method.

- **doublesex**
  locus (dsx) in Drosophila in which some mutations convert both males and females into intersexes. Encodes male and female specific proteins produced by differential splicing of the primary transcript in each sex.

- **doubling time**
  the average time required to double the number of individuals of a population.

- **downstream**
  a convention used to describe features of a DNA sequence, gene or mRNA related to the position and direction (5' to 3') of transcription by RNA polymerase or translation by the ribosome. Downstream (or 3' to) is in the direction of transcription (or translation) whereas upstream (5' to) is in the direction from which the polymerase (or ribosome) has come. Conventionally DNA sequences, gene maps and RNA sequences are drawn with transcription (or translation) from left to right and so downstream is towards the right.

- **drain**
  in order to remove water from the soil by artificial means (e.g., drainage ditches, buried perforated plastic pipes or a gravel sump.)

- **drainage**
  excess of water can be harmful to crop production, wet soils are usually low in temperature and low in oxygen content. Drainage can be facilitated by open ditches or different subsoil drains.

- **dressing**
  manure, compost or other fertilisers.

- **drift**
  changes in gene and genotypic frequencies in small populations due to random processes.

- **drill**
  a machine for seeding crops by dropping them in rows and

covering them with earth, the term is also used for a row of seeds deposited in the earth (i.e., the trench or channel in which the seeds are deposited), in general, to sow seeds in rows (i.e., the field was drilled, not sown broadcast.)

- **drill-row**
  drill.

- **drip irrigation**
  a system of watering by which moisture running through a porous hose, the water is slowly released through tiny holes or emitters to the plant roots, it is one of the most efficient of irrigation technologies.

- **drought hardening**
  adapting plants to survive periods of time with little or no water by stepwise reducing water supply or germinating and/or growing under insufficient moisture conditions.

- **drought stress**
  stress protein(s).

- **drought-tolerant**
  plants that can survive periods of time with little or no water.

- **drupe**
  a fleshy fruit, such as a plum, cherry, coconut, walnut, peach or olive, containing one or more seeds, each enclosed in a stony layer that is part of the fruit wall (hard endocarp.)

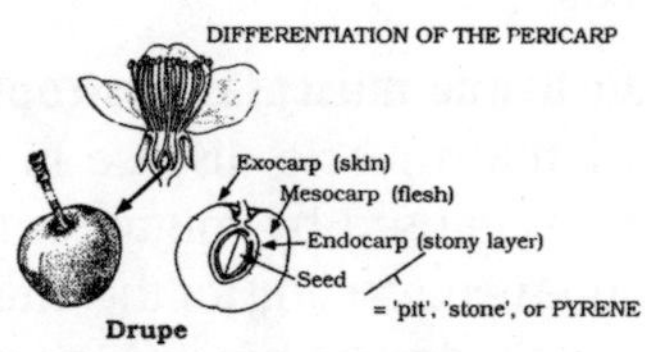

- **dry farming**
  a method of farming in arid and semiarid areas receiving less than 500 mm rainfall the year without using irrigation, the land being treated so as to conserve moisture. The technique consists of cultivating a given area in alternate years, allowing moisture to be stored in the fallow year. Moisture losses are reduced by producing a mulch and removal of weeds.

- **dry matter (DM)**
  the substance in a plant or plant material remaining after oven drying to a constant weight at a temperature slightly above the boiling point of water.

- **dry season**
  a period each year during which there is little precipitation.

- **dry weight**
  moisture-free weight.

- **duble helix**
structure of DNA consisting of two helices around a common axis.

- **duchenne muscular dystrophy**
a lethal muscle disease in humans caused by mutation in a huge gene coding for the muscle protein dystrophin, inherited as an X-linked recessive phenotype.

- **duplex type**
a polyploid plant that shows two dominant alleles for a given locus.

- **duplicate gene**
an identical locus mapping to a second site in a genome.

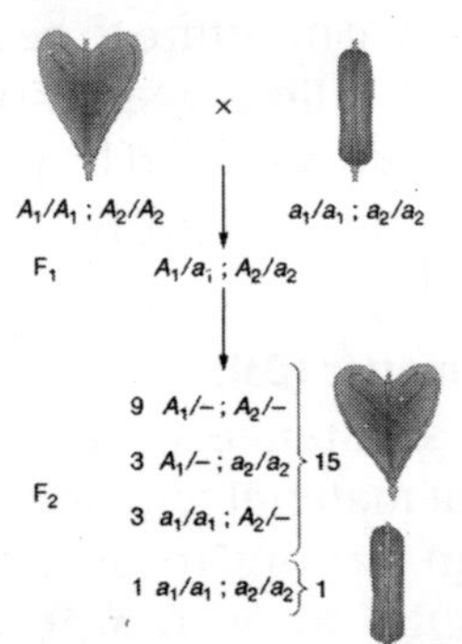

- **duplicate genes**
two or more pairs of genes in one diploid individual, which alone or together produce identical effects.

- **duplication**
more than one copy of a particular chromosomal segment in a chromosome set. Duplications supply genetic material capable of evolving new functions.

- **duplication of germplasm**
a duplicated seed sample that is prepared for safety reasons, usually the two samples are kept at different locations.

- **durable resistance**
resistance that remains effective during the agronomic life of a crop. This type of resistance is usually determined by several genes and is a main target of resistance breeding.

- **dust**
a method of pesticide control in which a dry substance is applied by spraying seeds, tubers or bulbs.

- **dust separator**
aspirator.

- **dwarf fruit trees**
a small fruit tree reaching a height at maturity of 150-200 cm, bred for convenient harvest technology as well as bearing early and normal fruits.

- **dwarfing gene**
one category of genes that control the height of the plant.

- **dyad**
two sister chromatids attached to the same centromere. A pair

of sister chromatids joined at the centromere, as in mitosis and meiosis.

- **dyeing flowers**
blossoms capable of producing dyes.

- **dysgenic**
eugenic.

- **dysploid**
a plant or species in which the chromosome number is more or less than the expected normal euploid number.

- **dysploidy**
abnormal ploidy (e.g., the appearance of diploid or triploid individuals in a normally tetraploid population or of triploid and tetraploid ones in a normally diploid population) See **anisoploid**

- **dystonia**
neurological condition involving repeated twisting and movement. Involves a variety of muscle groups. Intelligence not effected. Three forms: childhood - autosomal dominant, autosomal recessive, adult-acquired.

- **e (exit site)**
site on the ribosome through which tRNAs pass after they have donated their amino acid to the growing nascent polypeptide chain.

- **E coli**
common bacterium that has been studied intensively by geneticists because of its small genome size, normal lack of pathogenicity and ease of growth in the laboratory.

- **ear lifter (on a harvester)**
improves cutting efficiency with laid crops or overhanging ears.

- **early generation test**
selection schemes in which poor recombinants are discarded already in F2 and F3 generations.

- **ear-to-row planting**
See **ear-to-row selection.**

- **ear-to-row selection**
a separate growing of progenies (i.e., the separate sowing of lines or families), a procedure developed by the German breeder T. Roemer (*syn* Ohio method) Illinois method.

- **ecdysone**
a moulting hormone in insects.

- **echinate**
having sharply pointed spines, prickly.

- **ecological dominance**
the state in plant communities in which one or more species, by their size, number or coverage,

exert considerable influence or control over the other species.

- **ecological niche**
  the position occupied by a plant in its community with reference both to its utilisation of its environment and its required associations with other organisms.

- **ecology**
  the study of the interrelationships between individual organisms and between organisms and their environment.

- **Economic Trait Loci (ECL)**
  sites in the genome that determine characteristics of economic importance.

- **ecophene**
  the range of different phenotypes produced by one genotype within a certain environment.

- **ecospecies**
  a locally adapted species, it shows minor changes of morphology and physiology to another species which are related to a habitat and are genetically determined.

- **ecosystem**
  the complex of an ecological community together with a biological component of the environment, which function together as a system.

- **ecosystem resilience**
  the capacity of an ecosystem to withstand dramatic impacts. An ecosystem is the dynamic complex of micro-organisms, plants and animals including human communities and their nonliving environment, interacting as a functional unit.

- **ecotype**
  a locally adapted population of a widespread species. It shows minor changes of morphology and physiology that are related to a habitat and are genetically determined. The individuals of an ecotype are only uniform in traits that provide them with special adaptation to specific environments, all the other characters may vary. Ecotypes may be found in perennial clover, alfalfa, grasses or other forage crops.

- **ecovalence**
  this parameter is a quantitative measure for the evaluation of ecological adaptability.

- **ectoparasite**
  an external parasite opuntia.

- **ectopic expression**
  the occurrence of gene expression in a tissue in which it

is normally not expressed. Such ectopic expression can be caused by the juxtaposition of novel enhancer elements to a gene during genetic manipulation of transgenic organisms.

- **ectopic integration**
  in a transgenic organism, the insertion of an introduced gene (transgene) at a site other than its usual locus.

- **ectosite**
  See **ectoparasite.**

- **ectozoon**
  See **ectoparasite.**

- **edaphon**
  all (micro)organisms living in the soil close to the plant.

- **eddish**
  stubble.

- **edging**
  a row of plants set along the border of a plot or flower bed.

- **EDTA**
  ethylenediamine tetra acetic acid.

- **eelworm**
  in cereals, there are two species of eelworm attacking plants, the cereal cyst nematode *(Heterodera avenae)*, which can infest cereal crops and the oat stem eelworm *(Ditylenchus dipsaci)*, which attacks oats and rye, as well as several other crops and weeds.

- **effective breeding population (size)**
  in general, the size of population, which is adjusted mathematically to permit comparisons with others. More specifically, it is the size of an equivalent ideal population, which is expected to experience the same increase in homozygosity over time (i.e., drift) as the population in question. The ideal population is one in which mating is at random in the absence of selection and in which all individuals have the same expected contribution to the next generation.

- **effective seedling**
  any seedling that has survived in reasonable vigour for some arbitrary time and is so sited that it should make an effective contribution to the crop.

- **efficiency of plating**
  number of plaques formed by a phage lysate, it can be different for different hosts.

- **efflorescence**
  the time or state of flowering, in general, blooming or flowering.

- **electrical conductivity test**
  tests are used specifically for certain crops (e.g., pea), this test measures the soluble salts, sugars and amino acids that leach into the soak water. The electrical properties of the soak water are measured by a conductivity meter (expressed as micro siemans), the higher the leachate, the lower the vigour. This test is particularly useful in detecting mechanical and frost damages of cells.

- **electrical field fusion**
  See **cell fusion.**

- **electrofusion**
  See **cell fusion.**

- **electromorphs**
  allozymes that can be distinguished by electrophoresis.

- **electron microscope**
  a microscope that permits magnification of particles up to 200,000 diameters. Instead of having the specimen exposed to a light source, a stream of electrons is directed on the object. The higher resolving power of the electron microscope is largely the result of the shorter wavelength associated with electrons. The electrons are accelerated in a high vacuum through electromagnetic lenses and focussed on the specimen. They are projected on a fluorescent screen where the image of the particle may be viewed or onto a photographic plate and/or film.

- **electropermeabilisation**
  See **electroporation.**

- **electrophoresis**
  a technique for separating the components of a mixture of charged molecules (proteins, DNAs or RNAs) in an electric field within a gel or other support. The movement of electrically charged molecules in an electric field often results in their separation.

- **electroporation**
  a technique for transfecting cells by the application of a high-voltage electric pulse.

- **electrostatic separator**
  a machine that separates seed on the basis of their ability to accept and retain an electrical charge.

- **elicitor**
  a molecule produced by the pathogen host (or pathogen) that induces a response by the pathogen (or host).

- **elimination**
  culling.

- **elite**
  high-grade seed used for either seed and ware production, or, an agronomically superior and

high-performing local cultivar base seed.

- **elite germplasm**
  germplasm that is adapted (i.e., selectively bred and optimised to new environment). For example, maize, which is native to Mexico, has been adapted to many locations in the world.

- **elite hybrid variety**
  See **elite**.

- **elite plants**
  See **elite**.

- **elite population**
  genetically advanced intensively managed population in a short term breeding programme, sometimes used in same meaning as nucleus population.

- **elite seed**
  the class of pedigreed seed corresponding to the foundation seed class in some countries. See **base seed**.

- **elite strains**
  See **elite**.

- **elite tree**
  a tree that has been shown by progeny testing to produce superior offspring.

- **elongation factors (ef-ts, ef-tu, ef-g)**
  proteins necessary for the proper elongation and translocation processes during translation at the ribosome in prokaryotes. Replaced by eEF1 and eEF2 in eukaryotes.

- **eluate**
  a liquid solution resulting from eluting.

- **emasculate**
  to remove the anthers from a bud or flower before pollen is shed, a normal preliminary step in crossing to prevent self-pollination. There are basically two methods used for emasculation: individual emasculation and mass emasculation. Mass emasculation is applied, for example, in monoecious maize by mechanical detasseling, in hemp by removing male plants from the dioecious population, in rice by spike treatment with high temperatures (warm water of about +42°C), in wheat by using gametocides or in tobacco by dark-phase treatment during anthesis.

- **embl3 vector**
  a replacement vector for the cloning of large (~20 kb) DNA fragments, it derives from phage lamda.

- **embryo**
  the rudimentary plant within a seed that arises from the zygote, sometimes from an unfertilised

egg cell or from progressive differentiation in cell culture.

- **embryo culture**
  a method of inducing the artificial growth of embryos by excising the young embryos under septic conditions and placing them on sui nutrient media. The method is often applied when postgamous incompatibility exists (e.g., in wide crosses).

- **embryo percent**
  the amount of embryo compared with endosperm and other seed parts and/or the percent of embryo in the whole seed.

- **embryo rescue**
  See **embryo culture.**

- **embryo sac**
  the female gametophyte is formed by the division of the haploid megaspore nucleus—the site of fertilisation of the egg and development of the embryo.

- **embryogenesis**
  the formation of an embryo, after double fertilisation. Seed development begins with endosperm, embryo growth and differentiation, during subsequent seed maturation starch and lipids accumulate before the seed undergoes desiccation and dormancy. During late embryogenesis endosperm and mature embryo express seed storage reserve genes, many other hydrophyllic proteins with certain function that share similar sequence repeats also are accumulated. They are called late embryogenesis-abundant proteins, protecting cells during seed desiccation (e.g., abscisic acid).

- **embryogenic**
  related to or like an embryo.

- **embryoid**
  an embryolike structure.

- **embryonic (or tissue) polarity**
  the production of axes of asymmetry in a developing embryo or tissue primordium.

- **embryonic leaves**
  initial leaves, which are easily discernible in the germ of a mature grain.

- **emergence date**
  the time or date when plants start to emerge spikes, buds, flowers or showing coleoptiles, first leaves, etc.

- **empty fruits**
  in seed production, a fruit without a seed, for example, it happens in the composite family (sunflower, chicory, lettuce, etc.), the so-called seed is a fruit

(achene) of which the outer structures are comparable to the pod in the legumes, the true seed may or may not form inside the fruit, which is not (clearly) visible from outside.

- **enantiomer**
  mirror image form of a molecule. There may be more than two, as several parts of the molecule may have a mirror image form.

- **enation**
  an outgrowth from the surface of a leaf or other plant part.

- **encapsidation (of viruses)**
  the process when a virus forms the protein shell that surrounds the virus nucleic acid.

- **encapsulation**
  the process of enclosing fragile organic material in a protective casing, sometimes of a semisolid nature. It is used for planting or moving somatic embryos.

- **endangered species**
  a species on the verge of extinction.

- **endemic**
  confined to a given geographic region. Diseases continuously occurring in a particular area.

- **end-filling**
  conversion of a sticky end (5' overhang only) to a blunt end by enzymatic synthesis of the complement to the single-stranded overhang.

- **end-labelling**
  attaching a radioactive or non-radioactive label specifically to the ends of a DNA molecule.

- **endocarp**
  inner layer of the fruit wall (pericarp), it may be hard and stony (peach pit), membranous (apple core) or fleshy (orange pulp.)

- **endocytosis**
  the uptake of cellular material through the cell membrane by formation of a vacuole.

- **endodermis**
  a specialised tissue in the roots and stems of vascular plants, composed of a single layer of modified parenchyma cells forming the inner boundary of the cortex.

- **endoduplication**
  the doubling of the haploid chromosome complement owing to the failure of cell-wall formation.

- **endogamy**
  sexual reproduction in which the mating partners are more or less closely related.

- **endogenote**
  bacterial host chromosome. See **merozygote**.

- **endogenous rhythm**

a type of rhythmic plant response or growth capacity that is not affected by external stimuli.

- **endomitosis**

chromosomal replication without nuclear or cellular division that results in cells with many copies of the same chromosome.

- **endonuclease**

1. an enzyme that cleaves or hydrolyses phosphodiester bonds within a polynucleotide chain.
2. any enzyme that cuts DNA at specific sites corresponding to specific base sequences within a polynucleotide chain. Each endonuclease cuts at its own specific site. It is used to identify genomes, genotypes and chromosomes and it acts as a genetic marker.

- **endophyte**

(micro) organisms living inside host plants. Seedborne, non-pathogenic, fungal endophytes are commonly found in symbiotic relationships with many members of the cool-season grass subfamily *Pooideae*. There are beneficial effects on plants possessing fungal endophytes and detrimental effects on consumers of fungal endophyte-infected plants, e.g., fungi of the genus *Neotyphodium* were found in the diploid *Triticum* species (*T. dichasians*), a second endophyte, an *Acremonium* species, can be found in *T. columnare* , *T. cylindricum*, *T. monococcum* etc.

- **endoplasm**

the granular central material of the cytoplasm.

- **endoplasmic reticulum**

a system of minute tubules within the cytoplasm. Two types are recognised, rough and smooth, these are particularly concerned with pathways of protein and steroid synthesis.

- **endopolyploid**

1. diploid individuals whose cells containing 4C, 8C, 16C, 32C, etc. amounts of DNA in their nuclei.
2. an increase in the number of chromosome sets caused by replication without cell division.

- **endoreduplication**

chromosome reduplication in the interphase of the mitotic cell cycle.

- **endosperm**

in grasses, the reserve food material in the caryopsis lying outside the embryo. It is a starchy tissue that is formed during grain development, it provides nourishment for the developing embryo and for the seedling after germination until it can establish itself. Usually the endosperm is triploid and originates during the double fertilisation when one of the two sperm nuclei is fused with the two polar nuclei.

■ **endosperm texture**
the tendency of, for example, wheat endosperm to fracture either along the outlines of the cells (hard) or across the cells in a more irregular way (soft), it can be assessed by 'grinding time' or by inspection of the grain cross-section.

■ **endospory**
a condition in which the gametophyte develops within the spore wall, rather than externally.

■ **endotoxin**
a poison produced within a cell and released only when the cell disintegrates.

■ **enforced outbreeding**
deliberate avoidance of mating between relatives.

■ **engraft**
See **ingraft**.

■ **enhancer**
a modifier gene that may enhance the action of another gene.

■ **enhancer trap**
a transgenic construction inserted in a chromosome which is used to identify tissue-specific enhancers in the genome. In such a construct, a promoter sensitive to enhancer regulation is fused to a reporter gene, such that expression patterns of the reporter gene identify the spatial regulation conferred by nearby enhancers.

■ **enneaploid**
a polyploid plant with nine chromosome sets.

■ **enriched medium**
see **complete medium**.

■ **entomophilous**
insect-borne pollen, entomophilous pollen is usually characterised by a sticky, mainly liquid layer that fills in the interstices of the sculpted pollen wall or exine and is responsible for its adhesion to insect and other vectors. This liquid pollen coat, the tryphine, is also responsible for pollen adhesion to the stigmatic surface in many species.

■ **enucleate (of cells)**
removing the nucleus from a cell.

■ **enucleate cell**
a cell having no nucleus.

■ **environment**
1. the combination of all the conditions external to the genome that potentially affect its expression and its structure.
2. the sum of biotic and abiotic factors that surround and influence an organism.

■ **environmental mutagen**
a substance that may act as a mutagen in the environment of an organism.

■ **environmental resistance**
all characteristics that protect the reproduction of plants against negative influences of the environment.

■ **environmental variance**
the variance due to environmental variation.

■ **enzyme**
a protein that acts as a catalyst, speeding the rate at which a biochemical reaction proceeds but not altering the direction or nature of the reaction.

■ **epiallele(s)**
alternative states of a gene that have an identical DNA sequence but differ in methylation or chromatin structure and, hence, level of expression.

■ **epiblast**
the primordial outer layer of a young embryo.

■ **epicarp**
the outermost layer of a pericarp, as the rind or peel of certain fruits.

■ **epicotyl**
the portion of the embryo or seedling above the cotyledons.

■ **epidemic**
a rapid increase in disease over time and in a defined space.

■ **epidemic potential**
the biological capacity of a pathogen to cause disease in a particular environment.

■ **epidemic rate**
the amount of increase of disease in a plant population per unit of time.

■ **epigeal**
seed germination in which the cotyledons are raised above the ground by elongation of the hypocotyl (e.g., bean).

■ **epigeal germination**
it characterises a type of germination in which the cotyledons are raised above the ground by elongation of the hypocotyl.

- **epigenetic**
  reversible, nonhereditary variation that may be the result of changes in gene expression.

- **epigenetic regulation**
  regulation of gene activity mediated by a reversible change in DNA modification or chromatin structure.

- **epigenotype**
  See **epimutation.**

- **epiphytic**
  growing on other plants for its physical support but not drawing nourishment from them.

- **episome**
  a genetic element in bacteria that can replicate free in the cytoplasm or can be inserted into the main bacterial chromosome and replicate with the chromosome. Term used by Jacob and Wollman for genetic elements that can either exist independently in a cell or become integrated into the host chromosome.

- **episome**
  a bacterial plasmid that can integrate reversibly with the bacterial chromosome and replicate it.

- **episperm**
  seed coat.

- **epistasis**
  the nonreciprocal interaction of nonallelic genes, the situation in which one gene masks the expression of another. More specifically, gene interaction in which one gene interferes with the phenotypic expression of another nonallelic gene so that the phenotype is determined effectively by the former, the latter is described as hypostatic.

- **epistatic variance**
  variance epistasis.

- **epitope**
  a site on an antigen at which an antibody can bind, the molecular arrangement of the site determining the specific combining antibody.

- **epsp synthase**
  an enzyme in plants metabolic pathways that leads to aromatic amino acid production necessary for the development of proteins essential to plants growth.

- **equational division**
  the second meiotic division is an equational division because it does not reduce chromosome numbers. A nuclear division that maintains the same ploidy level of the cell.

- **equational separation**
  equational division.

- **equatorial plane**
  the plane between the two daughter nuclei of a dividing cell.

- **equatorial plate**
  an arrangement of the chromosomes in which they lie approximately in one plane, at the equator of the sp+indle. It is seen during metaphase of mitosis and meiosis.

- **equilateral**
  the type of panicle (e.g., in oats) in which the branches appear to spread equally on all sides.

- **equivalence group**
  a set of immature cells that all have the same developmental potential. In many cases cells of an equivalence group end up adopting different fates from one another.

- **eradicant**
  a chemical substance that destroys a pathogen at its source.

- **eradicate**
  to remove entirely, to pull up by the roots disease eradication.

- **erect**
  upright.

- **erectoides mutant**
  in cereals, a mutant form showing upright tillers and/or leaves.

- **error variance**
  variance arising from unrecognised or uncontrolled factors in an experiment with which the variance of recognised factors is compared in tests of significance.

- **escape**
  applied to a plant that has escaped from cultivation and naturalised more or less permanently. In plant pathology, the failure of inherently susceptible plants to become diseased, even though disease is prevalent.

- **espalier**
  a trelliswork of various forms on which the branches of fruit trees, grapevine, etc. are extended horizontally, in fan or other shape, in a single plane in order to provide better air circulation and sun exposure for the plants.

- **essential oils**
  the volatile, aromatic oils obtained by steam or hydrodistillation of botanicals. Most of them are primarily composed of terpenes and their oxygenated derivatives. Different parts of the plants can be used to obtain essential oils, including the flowers, leaves, seeds, roots, stems, bark and wood. Certain

cold-pressed oils, such as the oils from various citrus peels, are also considered to be essential oils, but these are not to be confused with cold-pressed fixed or carrier oils, such as olive, grapeseed and apricot kernel, which are nonvolatile oils composed mainly of fatty acid triglycerides.

- **establishment**
  the process of developing a crop to the stage at which the young plant may be considered established (i.e., safe from juvenile mortality and no longer in need of special protection).

- **establishment period**
  the time elapsing between the initiation of a new crop and its establishment.

- **ester**
  a compound that is formed as the condensation product, water removed, of an acid and an alcohol, while water is formed from the OH of the acid and H of the alcohol.

- **estivation**
  stagnating or otherwise nonfunctional during the summer period.

- **étagère**
  a series of open shelves for growing and displaying plants or in vitro cultures.

- **ethereal oil**
  See **essential oil.**

- **ethidium**
  a molecule that can intercalate into DNA double helices when the helix is under torsional stress.

- **Ethidium Bromide (EB)**
  fluorescent molecule that intercalates between base pairs of DNA and RNA.

- **Ethyl Methane Sulphonate (EMS)**
  a chemical compound that acts as a mutagen in plants mutagen.

- **ethylene**
  a volatile plant hormone, synthesis is promoted by auxin or damage in seedlings, in shoot apex and various organs. It is known as a ripening hormone. It stimulates flowering and fruit ripening, inhibits elongation of stems, roots and leaves.

- **ethylenediamine tetra acetic acid**
  known as edetic acid or tetra acetic acid, this chelator is a synthetic amino acid having a molecular weight of 292.25 and a molecular formula of $C_{10}H_{16}N_2O_8$. It is used in a wide range of biochemical and chemical procedures.

■ **etiolation**
a plant syndrome caused by suboptimal light, consisting of small, yellow leaves and abnormally long internodes.

■ **etiology**
the study of causes of diseases.

■ **e-type**
in sugarbeet breeding the high-yielding (E = Ertrag = yield) varieties with average sugar content.

■ **euapogamy**
a form of apomixis in which the sporophyte develops from a gametophyte without fertilisation and formation of zygotes.

■ **eucell**
a eukaryotic cell showing a nucleus, nuclear envelopes, chromosomes and nuclear divisions.

■ **eucentric**
a chromosomal interchange by which the translocated segment does not change the relative position to the centromere.

■ **euchromatin**
chromatin that shows the staining behaviour of the majority of the chromosome complement. It is uncoiled during interphase and condenses during mitosis, reaching a maximum density at metaphase.

■ **euchromatisation**
the induced or spontaneous change of heterochromatin into euchromatin.

■ **euchromosome**
a chromosome showing the typical features of the standard complement of a given autosome.

■ **eugenic**
favourable to the genetic quality of a population, as opposed to dysgenic.

■ **eugenics**
1. controlled human breeding based on notions of desirable and undesirable genotypes.
2. the improvement of humanity by altering its genetic composition by encouraging breeding of those presumed to have desirable genes.

■ **euhaploid**
a haploid genome showing no deviating number of chromosomes compared with the standard genome of the species.

■ **eukaryon**
the highly organised nucleus of a eukaryote.

■ **eukaryote**
cell or organism with membrane- bound, structurally dis-

crete *nucleus* and other well-developed subcellular compartments. Eukaryotes include all organisms except viruses, bacteria and blue- green algae. See **chromosomes**.

- **eukaryotic**
  eukaryote.

- **eukaryotic cell**
  a cell containing a nucleus.

- **euploid**
  a cell having any number of complete chromosome sets or an individual composed of such cells.

- **eupycnotic**
  normally coiled and normally stainable chromosomes.

- **eusexual**
  showing regular alternation of karyogamy.

- **eusom**
  a plant showing each member of the chromosome complement with the same copy number.

- **evaluation**
  the recording of those characters whose expression is often influenced by environmental factors.

- **everbloomer**
  a plant that blooms continuously throughout the growing season, for example, busy lizzie *(Impatiens walleriana)*.

- **evolution**
  1. in Darwinian terms a gradual change in phenotypic frequencies in a population that results in individuals with improved reproductive success.
  2. the process by which new species are formed from pre-existing species over a period of time.

- **evolutionarily conserved**
  see **conserved sequence**.

- **evolutionary**
  evolution.

- **evolutionary breeding**
  breeding procedure in which the variety is developed from an unselected progeny of a cross or multiple crosses, that have undergone evolutionary changes.

- **evolutionary divergence**
  the mode of evolutionary change whereby an ancestral population is split into different genotypic populations or phyletic lines.

- **evolutionary plasticity**
  the degree of genetic adaptability of a species or other taxa.

- **evolutionary rate**
  the rate of divergence between taxonomic groups measurable

as amino acid substitutions per million years.

- **ex situ**
  out of place or not in the *original* environment (e.g., seeds stored in a genebank.)

- **ex situ conservation**
  literally conservation 'off-site.' The conservation of a plant outside of its original or natural habitats, such as in a gene bank (a facility where temperature and humidity are artificially controlled) or botanical garden and stored as a seed, tissue, entire plant or pollen.

- **ex vito**
  describes plants or organs that are transplanted from culture to soil or to pots.

- **exchange pairing**
  the type of pairing of homologous chromosomes that allows genetic crossing-over to take place.

- **excised embryo test**
  a quick method for evaluating the growth potential of a root-shoot axis that has been detached from the remainder of the seed.

- **excision repair**
  a process whereby cells remove part of a damaged DNA strand and replace it through DNA synthesis using the undamaged strand as a template. The repair of a DNA lesion by removal of the faulty DNA segment and its replacement with a new segment.

- **exconjugant**
  each of the two cells that separates after conjugation has taken place. A female bacterial cell that has just been in conjugation with a male and that contains a fragment of male DNA.

- **exhauster**
  aspirator.

- **exhaustion test**
  a type of vigour test that measures the ability of seeds to grow rapidly under rigidly controlled conditions of high temperatures, relative humidity and moisture content in continuous darkness.

- **exine**
  the outer, decay-resistant coat of a pollen grain or spore, it shows different characteristics on the surface, which allows taxonomic differentiation between genera and even species. There are several scanning electron microscopy studies made on crop plants, also for comparisons in crop plant evolution.

■ **exocarp**
the pericarp or ovary wall of angiosperm fruits is composed of three different layers, the outer layer is the exocarp.

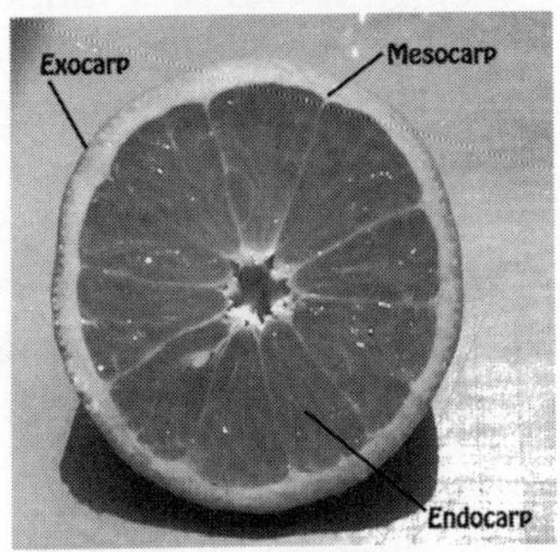

■ **exocytosis**
the extrusion of cellular material from a cell, as opposed to endocytosis.

■ **exogamy**
outbreeding, all forms of sexual reproduction in which the mating of unrelated and/or more distantly related partners is dominating.

■ **exogenote**
DNA that a bacterial cell has taken up through one of its sexual processes.

■ **exogenous DNA**
DNA originating outside an organism.

■ **exon**
a region of a gene that is present in the final functional transcript (mRNA) from that gene. Any non-intron section of the coding sequence of a gene, together the exons constitute the mRNA and are translated into protein.

■ **exon shuffling**
the hypothesis put forward by Walter Gilbert that exons code for functional units of a protein and that evolution of new genes has proceeded by recombination or exclusion of exons.

■ **exonuclease**
an enzyme that cleaves nucleotides one at a time from an end of a polynucleotide chain. An enzyme that hydrolyses phosphodiester bonds from either the 3' or 5' terminus of a polynucleotide molecules.

■ **exotic species**
a species that is not native to a region.

■ **exotoxin**
a poison excreted by a plant into the surrounding medium.

■ **experimental alteration of germplasm**
mutation.

■ **experimental design**
a branch of statistics that attempts to outline the way in which experiments should be carried out so the data gath-

ered will have statistical value.

- **experimental design**
  the planning of a process of data collection, also used to refer to the information necessary to describe the interrelationships within a set of data. It involves considerations, such as number of cases, sampling methods, identification of variables and their scale-types, identification of repeated measures and replications.

- **experimental lines**
  a group of individuals of a common ancestry and more narrowly defined than a strain or variety. A pure line is a clone. In plant breeding, 'line' refers to any group of genetically uniform individuals formed from a common parent.

- **explant**
  an excised fragment of a tissue or an organ used to initiate an in vitro culture.

- **explantation**
  explant.

- **exploitation competition**
  a form of competition that revolves around the superior ability to gather resources rather than an active interaction among organisms for these resources.

- **expressed gene**
  see *expression*.

- **Expressed Sequence Tags (ESTS)**
  DNA sequences derived by sequencing an end of a random cDNA clone from a library of interest. Usually, tens of thousands of such ESTs are generated as part of a given genome project. These ESTs provide a rapid way of identifying cDNAs of interest, based on their sequence 'tag'.

- **expression variegation**
  a type of variation in gene expression during development that causes streaks or patches of cells to have different phenotypes.

- **expression vector**
  vector constructs with a promoter sequence showing a highly efficient transcription of an inserted gene. The cell shows a high concentration of the gene product (i.e., protein).

- **expressivity of a gene**
  the degree to which a particular genotype is expressed in the phenotype.

- **extension**
  5' and 3' extensions are ssDNA regions at the ends of dsDNA, sometimes they are called overhangs or sticky ends.

- **extensograph**
a standard equipment for evaluating dough extensibility of cereal flour. Dough extensibility is an important quality parameter for biscuit making and thus a breeding target.

- **exterminate**
eradication.

- **extine**
See **exine.**

- **extra chromosome**
b chromosome.

- **extrachromosomal**
structures or processes outside the chromosomes.

- **extrachromosomal element**
all genetic elements that are not at all times part of a chromosome, for example, plasmids, phages, transposons, insertion sequences, plastid (mitochondrial and chloroplast) genomes.

- **extrachromosomal inheritance**
inheritance that is not controlled by chromosomal determinants but by cytoplasmic components.

- **extranuclear**
structures and processes outside the nucleus.

- **eye**
the centre of a flower when it is a different colour from the petals, the term is also used for an undeveloped bud on a tuber (e.g., potato) or for a cutting with a single bud.

- **eye depth (in potato)**
common scaling: 1 = very deep, 3 = deep, 5 = medium, 7 = shallow, 9 = very shallow.

- **eyepiece**
the lens or combination of lenses in an optical instrument (e.g., microscope) through which the eye views the image formed by the objective lens or lenses.

- **f-(minus) cell**
in *Escherichia coli*, a cell having no fertility factor, a female cell.

- **f'(prime) factor**
a fertility factor into which a portion of the bacterial chromosome has been incorporated.

- **f+(plus) cell**
in *Escherichia coli*, a cell having a free fertility factor, a male cell.

- **f1**
filial generation.

- **F1 generation**
the first filial generation, produced by crossing two parental lines.

- **F1 synthetic variety (syn-1)**
a variety derived by intercrossing a specific set of clones or seed-propagated lines, they

may include varieties of normally cross-fertilising or self-fertilising crops into which mechanisms have been introduced to maximise cross-fertilisation, e.g., Vitagraze' rye.

- **F2**
the progeny produced by intercrossing or self-fertilisation of F1 individuals.

- **F2 generation**
the second filial generation, produced by selfing or intercrossing the F1.

- **F2 variety**
it refers to the next generation seed derived from the hybrid (F1) generations, the variety cannot be perpetuated by growing additional generations, e.g., the tomato variety 'Foremost F2'.

- **F3**
progeny obtained by self-fertilising F2 individuals.

- **facilitated recurrent selection**
a type of recurrent selection in which genetic male sterility is maintained in the population to maintain heterozygosity and genetic diversity and to permit the recombination and shifting of gene frequencies.

- **factor**
synonymous with gene.

- **factor(ial) analysis**
a multivariate statistical analysis in which the independent variables are grouped into factors describing the variance of the dependent variables. It is used in cluster analysis.

- **factorial**
the product of all integers from the specified number down to one (unity).

- **factorial crossing group**
crossing group(s).

- **factorial design**
factorial trial.

- **factorial trial**
experimental design in which the effects of a number of different factors are investigated simultaneously. The factorial set of treatments consists of all combinations that can be formed from the different factors. The trial can include randomised complete blocks and Latin squares (e.g., treatments A and B at levels x and y), the sources of variance are replicates A, B, A x B and error.

- **facultative apomict**
apomicts that retain sexuality so that aberrant offspring may occur.

- **facultative growth habit**
  winter-and-spring wheat.

- **facultative heterochromatin**
  heterochromatin that is present in only one of a pair of homologues or not permanently present.

- **facultative parasite**
  a mainly saprophytic organism with weakly pathogenic properties.

- **facultative-type (of growth habit)**
  can be winter-and-spring type.

- **fading**
  photobleaching.

- **falcate**
  See **falciform.**

- **falciform**
  curved or sickle shaped (e.g., leaves or leaf hairs).

- **falling number (after hagberg)**
  the test provides an indication of alpha amylase activity and depends on the action of this enzyme in reducing the viscosity of a heated flour-and-water slurry. Alpha amylase is an enzyme involved in the degradation of starch to sugars and is usually associated with germination. Sprouted grains give low falling numbers, high number is required for better breadmaking quality (<200 low, 230 medium, >250 high, about 290 very high, >310 extremely high).

- **fallow (ground or cropland)**
  leaving the land uncropped for a period of time, it may contribute to moisture accumulation, to improvement of soil structure or to mineralisation of nutrients. It may be tilled or sprayed to control weeds and conserve moisture in the soil.

- **falls**
  the drooping or horizontal petals of irises.

- **false colour**
  representation in colours differing from the original scene.

- **false fruit**
  See **pseudocarp.**

- **false node**
  an abnormal node that occurs in some varieties of, for example, oats. A true node, appearing devoid of branches, occurs a little distance below it but the branches remain fused with the rachis and appear at the false node.

- **familial trait**
  a trait shared by members of a family.

- **family**
  a group of individuals directly related by descent from a

common ancestor, with at least one parent in common (one parent in common = half sibs, both parents in common = full sibs, selfing family = family obtained by self-pollinating a genotype).

- **family forestry**
tested open-pollinated, polyc-ross or full-sib families are deployed as single families to commercial plantations.

- **family selection**
1. a breeding technique of selecting a pair on the basis of the average performance of their progeny.
2. the selection of progeny families on their mean performance, in addition, the best individuals are usually selected in the best families.

- **fanning machine**
winnower.

- **fanning mill**
the air-screen machine that utilises airflow and sieving action in separating and cleaning seeds.

- **fan-shaped**
flabellate.

- **fan-training**
espalier.

- **farinogram**
a curve on a kymograph chart, it provides accessory information on dough properties, such as absorption, optimum mixing time and mixing tolerance. It may be used to estimate absorption and mixing time of a flour to be backed. Specific correlations between those traits have been established.

- **farmer's fixative**
three parts of anhydrous ethanol : one part of glacial acetic acid. a fixing and dehydrating agent used in histology, also used in conjunction with Feulgen stain and carbolfuchsin stain for chromosome analysis.

- **farmers' rights**
the recognition of farmers (past, present and future) as *in situ* agricultural innovators who collectively conserve and develop agricultural genetic resources around the world. As such, farmers are recognised as innovators entitled to intellectual integrity and to compensation whenever their innovations are commercialised.

- **farro**
it is known as hulled wheat, this means that the karyopsis retains its hull or husk during harvest and must be dehulled prior to

further processing. Its nutty flavour has long been popular in Europe, where it is also known as farro (Italy) and Dinkel (Germany). In Roman times it was called 'farrum', an ancient grain believed to have sustained the Roman legions, origins can be traced back to early Mesopotamia spelt (*Triticum spelta*) is an ancient and related to modern wheat (*Triticum aestivum*) it is one of the oldest of cultivated grains, preceded only by 'emmer' and 'einkorn'. In North America, this fine grain is commonly known as spelt, while there are occasional descriptions of spelt as not 'true' farro, the International Plant Genetic Resources Institute, via its report on Underutilised Mediterranean Species states that the only registered varieties of farro belong to spelt. It is an old Italian favourite (a famous wedding soup of these regions is called 'Confarrotio'). For centuries, farro has been a mainstay of Tuscany, Lazio, Umbria and Abruzzo in the northern part of Italy. These are relatively poor areas. Pushed aside in recent decades by easier to grow and harvest varieties of common wheat, farro is making a comeback among health conscious cooks and consumers. From a cross-country reading of the culinary winds, it appears that farro has finally made it to the New World. Used in soups, salads and desserts, the little light brown grain is an intriguing alternative to pasta and rice. Now farro (pronounced FAHR-oh) appears to be moving from rustic into fashionable restaurants not only in Tuscany and northern Italy (where it suddenly seems ubiquitous on menus), but also in the United States, particularly on the West and East Coasts wheat.

- **fasciata-type of pea**

a leaf mutant in peas, the leaves mutated into tendrils. It became a breeding target. Several varieties are commercially used.

- **fascicle**

a bundle of needles on a pine tree.

- **fasciculated root**

fibrous roots, in which some of the branches are thickened.

- **fat**

used by most plants as a energy-rich storage substance, usually present in the seeds. In some seeds it may amount about 70 percent of the dry matter. Plant fats are usually liquid at room temperature and they are mixtures of glycerine ester of many fatty acids.

- **fate map**
  a map of an embryo showing areas that are destined to develop into specific adult tissues and organs. A map of the developmental fate of a zygote or early embryo showing the adult organs that will develop from material at a given position on the zygote or early embryo.

- **father plant**
  the individual or species from which pollen was obtained to create a hybrid.

- **fatoid**
  a mutation that arises spontaneously in cultivated oats, the plant and grain closely resemble the variety in which it occurs but the grain shows to a varying degree certain characteristics of wild oats *(Avena fatua)*, for example, a strong geniculate awn, a 'horseshoe' base and dense hairs on the callus and the rachilla.

- **fatty acid**
  a long-chained, predominantly unbranched, carboxylic acid, in which a side-chain of carbon atoms is attached to the carboxyl group and hydrogen atoms to some or all of the carbon atoms in the side chain.

- **fecundity**
  the potential number of offspring produced during a unit of time resilience.

- **fecundity selection**
  the forces acting to cause one genotype to be more fertile than another genotype.

- **feed grain**
  any of several grains most commonly used for livestock feed, including maize, grain sorghum, soya, lupins, oats, rye, barley, etc.

- **feedback inhibition**
  a post-translational control mechanism in which the end product of a biochemical pathway inhibits the activity of the first enzyme of this pathway.

- **feedback mechanism**
  a control device in a system, homoeostatic systems have numerous negative-feedback mechanisms which tend to counterbalance positive changes and so maintain stability.

- **fen**
  low land covered wholly or partially with water.

- **ferment**
  enzyme.

- **fermentation**
  anaerobic respiration, usually applied to the formation of etha-

nol or lactate from carbohydrate biotechnology.

ferredoxin

a nonhaem iron protein with a low redox potential that functions as an electron carrier in both photosynthesis and nitrogen fixation.

fertile

a plant produces seed capable of germination or which produces viable gametes.

fertility factor (F factor)

the plasmid that allows a prokaryote to conjugate with and pass DNA into an F- cell. A bacterial episome whose presence confers donor ability (maleness).

fertilisation

the union of two gametes to produce a zygote that occurs during sexual reproduction.

fertiliser

a material that is added to the soil to supply one or more plant nutrients in a readily available form.

fertiliser grade

an expression that indicates the percentage of plant nutrients in a fertiliser.

festulolium

an artificial grass hybrid between *Festuca pratensis* x *Lolium perenne*, which is used in agriculture as a forage crop.

Feulgen reaction

a cytochemical test that utilises Schiff's reagent as a stain and DNA hydrolysis. It is highly specific for DNA detection. The method allows a wide range of chromosome studies and quantitative determination of DNA contents of nuclei applying the so-called cytophotometry. It was discovered in 1912 by Feulgen.

Feulgen staining

a histochemical and/or cytochemical reagent for quantifying nuclear DNA content or for staining chromosomes in cells, it is prepared as follows: 900 mL distilled water is boiled in 2-litre flask, then 5.0 g basic fuchsin is slowly added. It is swirled for one minute and then vacuum filtered through two layers of filter paper in a BUCHNER funnel into a 1-litre flask. It is then cooled to 50°C, while swirling 100 mL hydrochloric acid, 1.0 N 10.0 g potassium metabisulfite ($K_2S_2O_5$) is added and swirled for two minutes, the cloudy red solution will become clear blood red. It is stored in the refrigerator for twenty-four hours, then it is removed, warmed to room temperature and supplemented by 3.75 g activated charcoal. It is shaken vigorously for one minute and then vacuum filtered. The reagent should be clear and colourless, if not, the charcoal step must be repeated. The reagent has to be stored in the refrigerator and used within forty-eight hours, it remains for months.

- **few-seeded**
  oligospermous.

- **fibre**
  an elongated, thick-walled, often lignified cell (sclerenchyma) present in various plant tissues, usually providing mechanical support.

- **fibre crop**
  a crop plant mainly used for production of fibre (e.g., flax, abaca, hemp, etc.).

- **fibril**
  a small thread or very fine fibre, normally a fibre is constituted of a bundle of fibrils.

- **fibrillar**
  fibril.

- **fibrillarin**
  a nucleolar protein.

- **fibrous**
  resembling or having fibres.

- **fibrous root**
  a fine, densely branching root that absorbs moisture and nutrients from the soil.

- **ficoll**
  the brand name for an inert, synthetic, highly soluble polymer used as an osmotic agent, sometimes used for suspending protoplasts.

- **field burning**
  burning plant residue after harvest (1) to aid in insect, disease and weed control, (2) reduce cultivation problems and (3) stimulate subsequent regrowth and tillering of perennial crops.

- **field capacity**
  water that remains in soil after excess moisture has drained freely from the soil, usually expressed as a percentage of oven-dry weight of soil.

- **field crops**
  cash crops.

- **field diaphragm**
  a variable diaphragm located in the illumination pathway.

- **field experiment(ation)**
an evaluative test whereby the field performance of experimental plants is assessed in comparison to controls.

- **field gene bank**
a collection of accessions kept as plants in the field (e.g., perennial entries).

- **field germination**
a measure of the percentage of seeds in a given sample that germinate and produce a seedling under field conditions ground germination rate.

- **field grafting**
grafting a new variety on to an established rootstock already growing in the orchard.

- **field laboratory**
field test.

- **field moisture capacity**
the water that soil contains under field conditions field capacity.

- **field plane**
the set of plane(s) that are conjugate with the focused specimen. In a microscope adjusted for KOEHLER illumination it includes the planes of the specimen, the field diaphragm, the intermediate image plane and the image on the retina.

- **field plot size**
plot size.

- **field resistance**
synonymous to general resistance, it is under polygenic control (i.e., controlled by many genes with minor individual effects). In general, field resistance is longer lasting than race-specific resistance. Field resistance slows down the rate at which disease increases in the field.

- **field test**
an experiment conducted under regular field conditions (i.e., less subject to control than a precise contained experiment.)

- **field trial**
experiments carried out in the field field test.

- **filament**
the stalk of a stamen, which bears the anther.

- **filamentous**
threadlike filiform.

- **filial generation**
offspring generation. F1 is the first offspring or filial generation, F2 is the second and so on. Successive generations of progeny in a controlled series of crosses, starting with two specific parents (the P generation) and selfing or intercrossing the progeny of each new (F1, F2,.) generation.

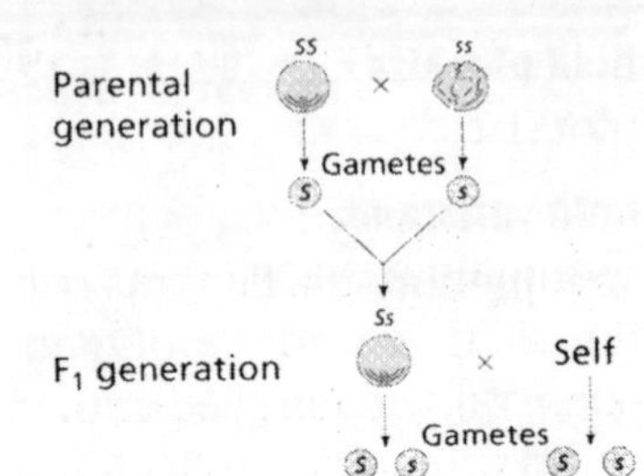

- **filial generation (f1)**
  the offspring resulting from first experimental crossing of the plants, the parental generation with which the genetic experiment starts is referred to as P1 F1.

- **filiation**
  descent.

- **filiform**
  threadlike, long and slender.

- **fill planting**
  the planting of plants in areas of inadequate stocking to achieve the desired level of stocking (density), either in plantations, areas of natural regeneration or other trials.

- **filling**
  beat(ting) up fill planting.

- **filter enrichment**
  a technique for recovering auxotrophic mutants in filamentous fungi (in which non-auxotrophic organisms are filtered off leaving a residue of non-growing auxotrophs.)

- **filter hybridisation**
  hybridisation of nucleic acid fragments on a filter (e.g., nitrocellulose) as a carrier SOUTHERN transfer.

- **final host**
  definite host.

- **firm seeds**
  sometimes applied to grass caryopses that are dormant due to seed coats that are impervious to water or gases.

- **First Division Segregation (FDS)**
  the allele arrangement (4+4) of spores within an ordered ascus that indicates the lack of recombination between a locus and its centromere. A linear pattern (4+4) of spore phenotypes within an ordered ascus for a particular allele pair, produced when the alleles go into separate nuclei at the first meiotic division, showing that no crossover has occurred between that allele pair and the centromere.

- **FISH**
  Florescent In Situ Hybridisation: a technique for uniquely identifying whole chromosomes or parts of chromosomes using florescent tagged DNA.

- **fishtail weeder**
  asparagus knife.

▪ **fissable**
fissile.

▪ **fissile**
See **fission.**

▪ **fission**
in genetics, the division of one cell by cleavage into two daughter cells or a chromosome into two arms.

▪ **fitness**
the relative ability of a plant to survive and transmit its genes to the next generation resilience.

▪ **fitness (w)**
the relative reproductive success of a genotype as measured by survival, fecundity or other life history parameters. See **Darwinian fitness and natural selection.**

▪ **fixation**
the first step in making permanent preparations of tissue, etc., for microscopic study. The procedure aims at killing cells and preventing subsequent decay with the least distortion of structure.

▪ **fixation agent**
a solution used for the preparation of tissue for cytological or histological studies. It precipitates the proteinaceous enzymes and prevents autolysis, destroys bacteria, etc. (e.g., acetic acid, formalin, Farmer's fixative, etc.).

▪ **fixative**
fixation agent.

▪ **fixed allele**
an allele for which all members of the population under study are homozygous, so that no other alleles for this locus exist in the population.

▪ **fixed breakage point**
according to the heteroduplex DNA recombination model, the point from which unwinding of the DNA double helices begins as a prelude to formation of heteroduplex DNA.

▪ **fixed effects model**
an effect of a treatment, in any experiment, which is concerned only with a certain, particular set of treatments rather than with the whole range of possible treatment effects.

▪ **flabellate**
fan-shaped.

▪ **flabelliform**
flabellate.

▪ **flag leaf**
the uppermost leaf on the grass stem and the last to emerge before the spike.

- **flanking region**
  the DNA sequences extending on either side of a specific locus or gene.
- **flash tape**
  a metalised plastic tape that produces bursts of light in response to breezes, it is suspended over crops to scare away birds.
- **flat**
  a container for holding packs of plant starter cells.
- **flexuous (spike neck)**
  wavy or in a more or less zigzag line.
- **flint maize**
  a variety of maize, *Zea mays* ssp. *indurata*, with very hard-skinned kernels.
- **floating check**
  check variety.
- **floating leaf**
  a leaf swimming on the surface of water (e.g., in lotus).
- **floating row cover**
  a fibre sheet, water and air penetrable, it is placed over a row or bed of plants for protection from heat, cold or insects.
- **flora**
  the plants of a particular region or period, listed by species and considered as a whole. In general, plants, as distinguished from fauna.
- **floral induction**
  the morphological changes in the development of a reproductive meristem from a vegetative meristem, it is the morphological expression of the induced state and usually occurs inside the meristem.
- **floral initiation**
  floral induction.
- **floral meristem**
  a meristem that gives rise to a flower.
- **floral primordium**
  floral meristem.
- **florescence**
  anthesis or flowering time or the state of being in bloom.
- **floret**
  a single flower consisting of the ovary, stamens and lodicules together with its enveloping lemma and palea in grasses.
- **floret initial**
  floral primordium.
- **floriculture**
  the cultivation of flowers or flowering plants.
- **floridean starch**
  an algal reserve resembling glycogen or amylopectin.

- **floriferous**
  bearing flowers.
- **florigen**
  the universal hormone that supposedly causes plants to change from the vegetative to the reproductive state.
- **flour**
  the finely ground meal of grain (e.g., from cereals) separated by bolting.
- **flour 'weak'**
  See **'strong' flour.**
- **floury**
  chalky.
- **flow cytometry**
  analysis of biological material by detection of the light- absorbing or fluorescing properties of cells or subcellular fractions (i.e., *chromosomes*) passing in a narrow stream through a laser beam. An absorbance or fluorescence profile of the sample is produced. Automated sorting devices, used to fractionate samples, sort successive droplets of the analysed stream into different fractions depending on the fluorescence emitted by each droplet.
- **flow densitometry**
  sorting.
- **flow karyotyping**
  use of flow cytometry to analyse and/or separate *chromosomes* on the basis of their DNA content.
- **flow sorting**
  sorting.
- **flower**
  a typical flower of angiosperms or plants whose seeds are enclosed in an ovary, it is composed of petals, sepals, stamens and a pistil, the flower morphology contributes to the relative importance of self- and cross-pollination. The two structures directly involved in sexual reproduction are the male stamina and the female pistil. A stamen consists of an anther, which contains the pollen grains and a filament on which the anther is borne.

- **flower bud initiation**
  floral primordium.
- **floweret**
  one of the segments of a cauliflower head.
- **flowering hormone**
  florigen.

fluctuation test
a test used in microbial genetics to establish the random nature of mutation or to measure mutation rates. An experiment by Luria and Delbruck that compared the variance in number of mutations among small cultures with subsamples of a large culture to determine the mechanism of inherited change in bacteria.

fluid drilling
a mechanical procedure for planting seed. Pregerminated seeds are suspended in a gel and sowed through a fluid drill seeder. This technology is potentially adaptive for sowing artificial seed, such as somatic embryos or embryoids.

fluorescence
property of certain molecules to absorb energy in the form of light and then release this energy at a longer wavelength than the wavelength of absorption (i.e., at a lower energy level.)

fluorescence in situ hybridisation (FISH)
a technique for visual detection in the microscope of specific DNA sequences on cytological fixed chromosomes, after hybridisation with DNA probes labelled with a fluorochrome. It can be done on both interphase and metaphase chromosomes chromosome painting.

fluorescence microscopy
the common method of microscopic examination based on observing the specimen in the light transmitted or reflected by it. Fluorescence preparations are self-luminous, the tissue is stained by fluorochromes, dyes that emit light of longer wavelength when exposed to blue or ultraviolet light, the fluorescing parts of the stained object then appear bright against a dark background the staining technique is extremely sensitive.

fluorescence staining
very few biological samples are inherently fluorescent such that they can be imaged directly, instead, fluorescence microscopy essentially always uses a fluorochrome that is introduced through some form of staining procedure. DAPI is a useful tool in various cytochemical investigations when DAPI binds to DNA, its fluorescence is strongly enhanced, which has been interpreted in terms of a highly energetic and intercalative type of interaction, but there is also evidence that DAPI binds to

the minor groove, stabilised by hydrogen bonds between DAPI and acceptor groups of AT, AU and IC base pairs other applications use single-stranded DNA sequences with a fluorescent label to hybridise with its complementary target sequence in the chromosomes, allowing it to be visualised under ultraviolet light. Fluorescence staining methods offer several advantages, such as high resolution, live cell staining and the possibility of dual or multiple labelling.

- **fluorescent**
  the colour exhibited when the grain or glumes of certain oat varieties are viewed under ultraviolet light or other radiation.

- **fluorescin**
  a red crystalline compound, $C_{20}H_{12}O_5$, which in alkaline solutions produces an intense green.

- **Fluorescin Diacetate (FDA) staining**
  living cells stained with FDA fluorescence in the presence of UV light. The stain is used to assess cell viability of cell cultures, etc.

- **fluorochrome**
  fluorescence dye.

- **flush cut**
  a pruning cut to remove a tree limb in which the cut is completely flush with the tree, the resulting scar is usually too large to heal efficiently.

- **flush ends**
  blunt ends.

- **flush season**
  the plant growth that is produced during a short period.

- **flux**
  a flowing or flow.

- **focal plane**
  a plane through a focal point and perpendicular to the axis of a lens, mirror or other optical system.

- **focus**
  the ability of a lens to converge light rays to a single point.

- **focus map**
  a fate map of areas of the Drosophila blastoderm destined to become specific adult structures, based on the frequencies of specific kinds of mosaics.

- **fodder**
  harvested grass or other crop parts for animal feed.

- **Fokker-Planck equation**
  an equation that describes diffusion processes. It is used by

population geneticists to describe random genetic drift.

- **foliaceous**
leaflike shape.

- **foliage**
the leaves of plants.

- **foliage blight**
late blight.

- **foliar feeding**
feeding plants by spraying liquid fertiliser on the leaves.

- **foliar nutrient**
any liquid substance applied directly to the foliage of a growing plant for the purpose of delivering an essential nutrient in an immediately available form.

- **foliar treatment**
treating plants by spraying liquid or dry insecticides, pesticides or herbicides on their leaves.

- **foliation**
leafing.

- **foliole**
leaflet.

- **foliose**
leafy.

- **follicle**
a fruit with a simple pistil that at maturity splits open along one suture.

- **follicular fruit**
follicle.

- **food grain**
cereal seeds most commonly used for human food, chiefly wheat and rice.

- **food legume**
legume plants with nutritive value for humans, directly and indirectly consumed.

- **food plant**
plants with nutritive value for humans, directly and indirectly consumed.

- **food species**
food plant.

- **foot rot**
eyespot disease.

- **footprinting**
a method used to determine the length of nucleotide chains that are close to a protein (which bind to DNA), for example, certain types of drugs act by binding tightly to certain DNA molecules in specific locations.

- **forage**
feed from plants for livestock such as hay, pasturage, straw, silage or browse.

- **forage crop**
crop plants for feeding of livestock (e.g., alfalfa, clover, maize) See **forage**.

- **forage shrub**
forage.

- **forceps**
pincers.

- **forcing**
the practice of bringing a plant into growth or flower (usually by artificial heat or controlling daylight) at a season earlier than its natural one. It is sometimes applied in order to synchronise flowering dates of parental plants for crosses in the greenhouse.

- **forecrop**
the crop grown during the season of the respectively present cropping.

- **foreign DNA**
DNA that is not found in the normal genome concerned. Usually it is directly or indirectly introduced into a recipient cell by several experimental means.

- **forest tree breeding**
the genetic manipulation of trees, usually involving crossing, selection, testing and controlled mating, to solve some specific problem or to produce a specially desired product.

- **forked**
sometimes a morphological deviation of the common root in beets (e.g., sugarbeet) fangy root.

- **form**
a botanical category ranking below a variety and differing only trivially from other related forms (e.g., in waxiness of leaves.)

- **forma specials (F.SP.)**
a taxon characterised from a physiological standpoint (especially host adaptation), in general, biotypes of pathogen species that can infect only plants that are within a certain host genus or species .

- **formaldehyde**
a colourless gas readily soluble in water.

- **formalin**
an aqueous solution of formaldehyde commonly used as a fixative that functions through cross-linking protein molecules.

- **formazan**
colourless when dissolved in water, the chemical 2,3,5-triphenyl tetrazolium chloride is reduced to the red-coloured chemical triphenyl formazan on contact with living, respiring tissue, the amount of formazan formed is used as a measure of seed viability, as it reflects oxidative metabolism tetrazolium test.

- **formylmethionine (FMET)**
a specialised amino acid that is the very first one incorporated

into the polypeptide chain in the synthesis of proteins in prokaryotes.

- **forward mutation**
  a mutation that alters (usually inactivates) a wild-type allele of a gene back mutation.

- **forward selection**
  choosing good individuals out of a progeny test for possible use in seed orchards and/or subsequent generations of breeding.

- **fossil**
  markedly outdated.

- **foundation seed**
  seed stocks increased from breeder seed, handled as to closely maintain the genetic identity and purity of the variety. It is a sort of certified seed, either directly or through registered seed.

- **foundation single cross**
  it refers to a single cross in the production of a double, three-way or top cross.

- **founder effect**
  genetic drift due to the founding of a population by a small number of individuals.

- **founder population**
  the first generation of a breeding population, e.g., in forest tree breeding often the initial plus trees, this is usually the starting point of calculations base population.

- **founders**
  founder population.

- **four-way cross**
  double cross.

- **F-pili**
  sex pili. Hair-like projections on an F+, an F' or Hfr bacterium involved in anchorage during conjugation and presumably through which DNA passes.

- **fragile sites**
  a non-staining gap of variable width that usually involves both chromatids and is always at exactly the same point on a specific chromosome derived from an individual or kindred.

- **fragile-X syndrome**
  the most common form of inherited mental retardation. Named for its association with an X chromosome with a tip that breaks or appears unconde-nsed. Inheritance involves imprinting.

- **frame shift mutation**
  a mutation that is caused by a shift of the reading frame of the mRNA (usually by the insertion of a nucleotide) synthesised from the altered DNA template.

- **frameshift**
  a mutation in which there is an addition or deletion of

one, two or a small number (not a multiple of three) of nucleotides that causes the codon reading frame to shift to one of two others from the point of the mutation during translation. Consequently the amino acid sequence of the protein is altered from the point of the mutation to the carboxy terminus.

- **frameshift mutation**
  the insertion or deletion of a nucleotide pair or pairs, causing a disruption of the translational reading frame.

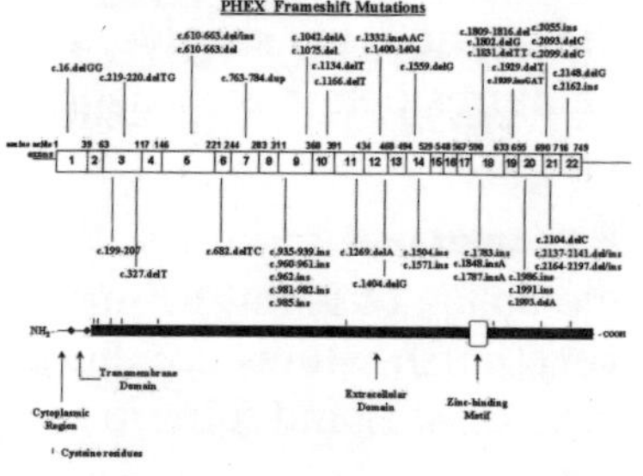

- **free nuclear division**
  it refers to mitotic division of nuclei without accompanying cytokinesis, i.e., nuclei divide in a common cytoplasm and the cells walls only forming around each later.

- **free-threshing**
  spikes with brittle (fragile) rachis, for example, in wheat, it is controlled by a single recessive allele. The fragility of the spike reveals the main difference between wild and cultivated forms.

- **freeze preservation**
  cryopreservation.

- **freezing injury**
  a type of winter injury caused by the combined effects of low temperature, wind and insufficient soil moisture. The low temperature injury is associated with ice formation in the extracellular spaces resulting in freeze-induced dehydration and metabolic changes, the plasma membrane remains attached to the cell wall, causing the cell to collapse.

- **frego-bract mutant**
  a mutant bract type in cotton in which the bracts curl outward, exposing flower buds and bolls.

- **frequency**
  a way of summarising a set of data, a record of how often each value (or set of values) of the variable in question occurs, may be enhanced by the addition of percentages that fall into each category. Used to summarise categorical, nominal and ordinal data, may also be used to summarise continuous data once the data set has

been divided up into sensible groups.

- **frequency histogram**
  a step curve in which the frequencies of various arbitrarily bounded classes are graphed.

- **frequency-dependent fitness**
  fitness differences whose intensity changes with changes in the relative frequency of genotypes in the population.

- **frequency-dependent selection**
  selection where the fitness of a type varies with its frequency (i.e., whereby a genotype is at an advantage when rare and at a disadvantage when common).

- **frequency-interdependent fitness**
  fitness that is not dependent upon interactions with other individuals of the same species.

- **friabilin(s)**
  proteins that determine the adhesion of the starch granules, for example, in wheat they determine the difference between hard and soft wheat, soft wheat's show strong friabilins, which bind the granules and hence the endosperm fractionates into large fragments, whilst hard wheat's contain weaker friabilins and hence fracture into small fragments. Latest molecular and biochemical evidences associate them with differences in the structure of proteins (puroindolines).

- **frond**
  the leaves of ferns and other cryptogams, includes both stipe and blade, commonly used to designate any fernlike or featherlike foliage.

- **frost damage**
  See **killing frost**.

- **frost killing**
  See **killing frost**.

- **frost resistance**
  the capacity to survive temperatures below zero degrees Celsius.

- **frost tolerance**
  the ability of plants to survive very harsh winter conditions (i.e., to withstand subzero temperatures).

- **frost-lifting of seedlings**
  heaving.

- **fructification**
  the process of forming a fruit body or the fruit body itself.

- **fruiting body**
  in fungi, the organ in which meiosis occurs and sexual spores are produced.

■ **functional alleles**
mutants that fail to complement each other in a cis-trans complementation test.

■ **fundamental number**
the number of chromosome arms in a somatic cell of a particular species.

■ **G**
guanine or guanosine.

■ **G banding**
a special staining technique for chromosomes that results in a longitudinal differentiation by Giemsa stain, which is a complex of stains specific for the phosphate groups of DNA, the characteristic bands produced are called G bands. These bands are generally produced in AT-rich heterochromatic regions differential staining.

■ **gain-of-function dominant**
a mutation in which dominance is caused by changing the specificity or expression pattern of a gene or gene product, rather than simply by reducing or eliminating the normal activity of that gene or gene product.

■ **galactose (gal)**
a component of milk sugar (lactose).

■ **gamet precision divider**
a type of mechanical halving device for subdividing a large seed sample to obtain a smaller working sample for germination or purity analysis. It has an electrically operated rotating cup into which the seed is funnelled to be spun out and into one of two spouts.

■ **gamete**
a germ cell having a haploid chromosome complement. Gametes from parents of opposite sexes fuse to form zygotes. A specialised haploid cell that fuses with a gamete from the opposite sex or mating type to form a diploid zygote, in mammals, an egg or a sperm.

■ **gametic selection**
the forces acting to cause differential reproductive success of one allele over another in a heterozygote.

■ **gametocidal**
See **gametocide.**

■ **gametocidal gene**
a gene encoding a product that destroys cells that divide to produce the gametes.

■ **gametocide**
a chemical agent used to selectively kill either male or

female gametes, it is used in hybrid seed production of autogamous crops (e.g., barley or wheat.)

- **gametoclonal**
regenerated from a tissue culture originating from gametic cells or tissue.

- **gametoclonal variation**
variation among regenerants obtained from pollen and/or anther culture.

- **gametoclone**
plants regenerated from cell culture derived from meiocytes or gametes.

- **gametocyte**
a cell that will undergo meiosis to form gametes.

- **gametogamy**
the fusion of the sexual gametes and the formation of the zygotic nucleus.

- **gametogenesis**
the formation of gametes from gametocytes.

- **gametophyte**
a haploid phase of the life cycle of plants during which gametes are produced by mitosis, it arises from a haploid spore produced by meiosis from a diploid sporophyte.

- **gametophytic apomixis**
agamic seed formation in which the embryo sac arises from an unreduced initial, it includes both diplospory and apospory.

- **gametophytic self-incompatibility**
self-incompatibility is based on the genotypic and phenotypic relationship between the female and male reproductive system. Alleles in cells of the pistil determine its receptivity to pollen. The phenotype of the pollen, expressed as its inability to effect fertilisation, may be determined by its own alleles, referred to as gametophytic incompatibility.

- **gamopetalous**
having the petals of the corolla more or less united.

- **gangrene of potato**
necrosis or death of soft tissue due to obstructed circulation, usually followed by decomposition and putrefaction (*Phoma exigua* var. *foveata*).

- **gap**
single-stranded region in dsDNA.

- **gapping**
beat(ting) up.

■ **gas chromatograph**
an analytical technique for identifying the molecular composition and concentrations of various chemicals in agricultural products, water or soil samples.

■ **gastrulation**
the process of movements and infoldings of embryonic cells destined to become endoderm in early animal embryos, immediately following blastula (or blastoderm) stage, generating the blastopore.

■ **Gauss distribution**
See **normal distribution.**

■ **Gaussian curve**
See **normal distribution.**

■ **GCA**
combining ability (CA) general combining ability.

■ **geitonogamy**
when neighbouring flowers of the same plant can achieve pollination, as opposed to xenogamy.

■ **gelatin(e)**
a nearly transparent, glutinous substance, obtained by boiling the bones, ligaments, etc. of animals, used in making jellies.

■ **GEM**
Genetically Engineered Microorganism.

■ **geminivirus**
a single-stranded DNA virus that causes serious diseases in cereals, vegetables and fibre crops worldwide. Like nanoviruses, it is transmitted by either whiteflies or leafhoppers. Whitefly-transmitted geminiviruses are all in the genus *Begomovirus,* which typically have bipartite genomes (DNA A and DNA B) comprising circular DNAs of ~2,700 nucleotides, although some have been shown to be monopartite, having only a DNA A component. In contrast, the nanoviruses are a recently established group of plant viruses that are transmitted by either aphids or planthoppers and have multipartite genomes comprising circular DNAs of ~1,000 nucleotides.

■ **gemmiferous**
bearing buds.

■ **gene**
the fundamental physical and functional unit of heredity. It is an ordered sequence of *nucleotides* located in a particular position on a particular *chromosome* that encodes a specific functional product (i.e., a *protein* or *RNA molecule*). See **expression.**

■ **gene action**
the expression of the gene based on the transcription into comple-

mentary RNA sequences, the subsequent translation of mRNA into polypeptides, which may form a specific protein.

- **gene activation**
  different mechanisms of repressing and activation of genes.

- **gene amplification**
  a process by which the cell increases the number of a particular gene within the genome. The process by which the number of copies of a chromosomal segment is increased in a cell.

- **gene bank**
  an establishment in which both somatic and hereditary genetic material are conserved (seeds, pollen, whole plants, extracted DNA). It stores in a viable form material from plants that are in danger of extinction in the wild and cultivars that are not currently in popular use. The stored genetic material can be called up when required. The normal method of storage is to reduce the water content of seed material to around 4 percent and keep it at 0°C or less (–20°C), all stored stocks are periodically checked by germination tests.

- **gene centre**
  it refers to the centre of origin of a given crop plant.

- **gene cloning**
  insertion of a DNA fragment carrying a gene into a cloning vector. Subsequent propagation of the recombinant DNA molecule in a host organism results in many identical copies of the gene (clones) in a form that is more easily accessible than the original chromosomal copy.

- **gene conservation population**
  a population used to maintain original genetic variation in species.

- **gene construction**
  an experimentally engineered gene with functional and non-functional properties.

- **gene conversion**
  a process whereby one member of a gene family acts as a blueprint for the correction of the other. This can result in either the suppression of a new mutation or its lateral spread in the genome.

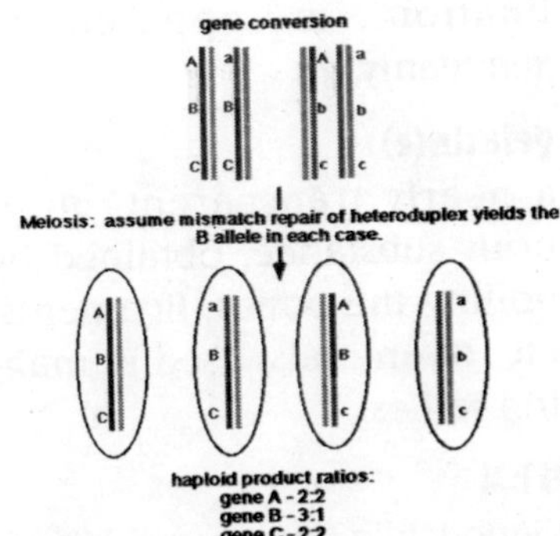

gene dosage (dose)
the number of times a given gene is present in the nucleus of a cell and/or individual.

gene duplication
a process in evolution in which a gene is copied twice, the two copies lie side by side along the same chromosome.

gene expression
the process by which a gene coded information is converted into the structures present and operating in the cell. Expressed genes include those that are transcribed into mRNA and then translated into protein and those that are transcribed into RNA but not translated into protein (e.g., transfer and ribosomal RNAs).

gene families
groups of closely related genes that make similar products.

gene flow
the spread of new genes, which takes place within an interbreeding group as a result of crossing with immigrants.

gene frequency
the number of loci at which a particular allele is found divided by the total number of loci at which it could occur for a given population, expressed as a proportion or percentage.

gene fusion
the accidental joining of DNA of two genes, such as can occur in a translocation or inversion. Gene fusions can give rise to hybrid proteins or to the misregulation of the transcription of one gene by the cis regulatory elements (enhancers) of another.

gene insertion
the addition of one or more copies of a normal gene into a defective chromosome.

gene interaction
modification of gene action by a nonallelic gene or genes, generally the interaction between products of nonallelic genes.

gene library
in molecular genetics, a random collection of cloned DNA fragments in a number of vectors that ideally includes all genetic information of that species.

gene linkage
linkage.

gene location
determination of physical or relative distances of a gene on a particular chromosome.

gene locus
the fixed position that a gene occupies on a chromosome.

- **gene machine**
in common literature, an idiomatic description of an automated oligonucleotide synthesiser.

- **gene map**
1. a linear designation of mutant sites within a gene, based upon the various frequencies of interallelic (intragenic) recombination.
2. the DNA sequence of a gene annotated with sites of regulatory elements, introns, exons and mutations.

- **gene mapping**
determination of the relative positions of genes on a DNA molecule (chromosome or plasmid) and of the distance, in linkage units or physical units, between them.

- **gene mutation**
a heritable change of gene revealed by phenotypic modifications.

- **gene pairs**
the two copies of a gene present in a diploid, one on each homologous chromosome.

- **gene patenting**
protection provided by governmental or nongovernmental institutions to the discoverer of new genes, genotypes, strains or testing procedures so that the detailed information can be declared publicly. A synthetic gene, but not a natural gene itself, can be patented, however, its sequenced functional unit or specific utilisation can be a matter of patent.

- **gene pool**
the reservoir of different genes of a certain plant species or lower and higher taxa available for crossing and selection, it may be differentiated between (1) primary gene pools (consists of those species that readily hybridise, produce viable hybrids and have chromosomes that may freely recombine), (2) secondary gene pools (consists of those species with a certain degree of hybridisation barrier due to ploidy differences, chromosome alterations or incompatibility genes) and (3) tertiary gene pools (consists of distinct species or higher taxa with strong crossing barriers). In general, the total number of genes or the amount of genetic information that is possessed by all the reproductive members of a population of sexually reproducing organisms.

- **gene product**
the biochemical material, either RNA or protein, resulting from expression of a gene. The

amount of gene product is used to measure how active a gene is, abnormal amounts can be correlated with disease- causing alleles.

- **gene recombination**
recombination.

- **gene redundancy**
the presence of a gene(s) in multiple copies due to polyploidy, polytenic chromosomes, gene amplification or chromosomal duplications.

- **gene splicing**
combining genes from different organisms into one organism.

- **gene stacking**
the insertion of two or more (possibly synthetic) genes into the genome (e.g., the *bat* gene from *Bacillus thuringiensis* and a gene for resistance to a specific herbicide,)

- **gene substitution**
the replacement of one allele by another mutant allele in a population by natural or directed selection.

- **gene symbol**
designating a gene, usually by an abbreviation of the name or description of a given gene. In the past, genes have been described by Latin names and subsequently the one- to three-letter abbreviations. Currently, several systems of naming and symbolisation are in use, although a comparable, uniform symbolisation is sought.

- **gene tagging**
the labelling of a gene by a marker gene or specific DNA sequence closely linked with the gene in question.

- **gene targeting**
the insertion of antisense DNA molecules in vivo into selected cells of the body in order to block the activity of undesirable genes. These genes might include oncogenes or genes crucial to the life cycle of parasites.

- **gene technology**
in a broad sense, it is the artificial transfer of genes between cells or individuals by means of molecular and in vitro techniques, prerequisites are (1) the presence of a gene, which is available as a DNA fragment, (2) a clonable DNA, (3) the fragment has to be transferable by different systems, (4) the incorporation of the DNA fragment into a recipient cellular genome has to be feasible, (5) the transformed cells have to be regenerable into a normal plant and (6) the gene that was transferred

must be expressed in the alien genetic background.

- **gene therapy**
  addition of a functional gene or group of genes to a cell by gene insertion to correct an hereditary disease.

- **gene transfer**
  the physical transfer of a gene by crossing, chromosomal manipulation and molecular means. In biotechnology, different methods are described, such as (1) microinjection, (2) insertion via microprojectiles (particle gun, particle bombardment) using silicon fibres as carriers of the DNA, (3) direct transfer or (4) a vector-mediated transfer.

- **gene translocation**
  the transfer or movement of a gene or gene fragment from one chromosomal location to another, often it alters or abolishes expression position (positional) effect.

- **genealogy**
  descent from an original form or progenitor, ancestry.

- **gene-for-gene theory**
  in certain plant-pathogen interactions, a gene for resistance in the host corresponds to and is directed against a gene for virulence in the pathogen.

- **genepool system**
  it consists of three informal categories in order to provide a genetic perspective and focus for cultivated plants.

- **general combining ability**
  combining ability (CA).

- **general resistance**
  resistance against all biotypes of a pathogen, nonspecific host-plant resistance.

- **general seed blower**
  a precision seed blower used to aid in separating light seed and inert matter from heavy seed.

- **generalised resistance**
  general resistance.

- **generalised transduction**
  the ability of certain phages to transduce any gene in the bacterial chromosome. Form of transduction in which any region of the host genome can be transduced.

- **generation time**
  the time required for a culture to double its cell number.

- **generative**
  sexual processes.

- **generative meristem**
  gives rise to parts, such as floral organs that ultimately produce fruits and seeds.

■ **generative nucleus**
a haploid nucleus of a pollen grain that produces two sperm nuclei by mitosis (pollen grain mitosis) sperm nucleus.

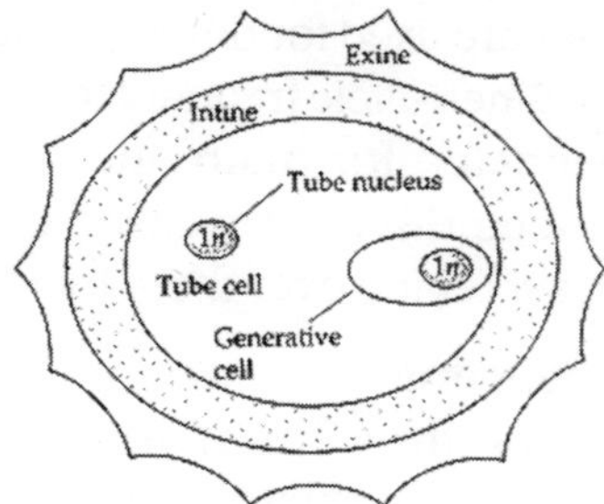

■ **generic**
referring to the genus.

■ **generic diversity**
the differences between individuals of different genera.

■ **genetic**
pertaining to the origin or common ancestor or ancestral type.

■ **genetic (linkage) map**
the linear arrangement of gene loci on a chromosome, deduced from genetic recombination experiments, a genetic map unit is defined as the distance between gene pairs for which one product of meiosis out of a hundred is recombinant (i.e., it equals a recombination frequency of 1 percent.)

■ **genetic advance**
the expected gain in the mean of a population for a particular quantitative character by one generation of selection of a specified percent of the highest-ranking plants.

■ **genetic architecture**
the distribution of genetic variation in a species, usually described hierarchically as variation at the regional, local, family and individual levels and also relating to proportions of additive and non-additive inheritance.

■ **genetic background**
the remaining genetic constitution when a particular locus or allele of a given individual or taxon is studied.

■ **genetic balance**
the optimal interaction of co-adapted genes, alleles or genetic systems within a given individual.

■ **genetic block**
the reduction or stop of an enzyme activity caused by a specific gene mutation.

■ **genetic code**
linear sequences of three nucleotides (triplets) that specify amino acids or termination (nonsense) codons during the

process of translation at the ribosome. The correspondence between nucleotide triplets in DNA and amino acids in protein.

- **genetic complement**
genome.

- **genetic complementation**
the complementary action of homologous sets of genomes.

- **genetic correlation**
the correlation between the genotypic values of two characters with respect to the genetic character.

- **genetic death**
death of an individual without reproducing, caused by mutationally arisen alleles that reduce the fitness of a genotype and/or taxon.

- **genetic dissection**
the use of mutation and recombination to piece together the various components of a given biological function.

- **genetic distance**
a measure of gene differences between individuals or populations measured by the differences of several characters. Such distance may be based on phenotypic traits, allele frequencies or DNA sequences (e.g., genetic distance between two populations having the same allele frequencies at a particular locus and based solely on that locus is zero). The distance for one locus is maximum when the two populations are fixed for different alleles. When allele frequencies are estimated for many loci, the genetic distance is calculated by averaging over these loci.

- **genetic drift**
the random fluctuations of gene frequencies in a population such that the genes amongst offspring are not a perfectly representative sampling of the parental genes.

- **genetic engineering**
the manipulation of DNA using restriction enzymes, which can split the DNA molecule and then rejoin it to form a hybrid molecule—a new combination of nonhomologous DNA. The technique allows the bypassing of all the biological restraints to genetic exchange and mixing and even permits the combination of genes from widely differing species.

- **genetic engineering technologies**
see **recombinant DNA technologies.**

■ **genetic equilibrium**
an equilibrium in which the frequencies of two alleles at a given locus are maintained at the same values generation after generation. A tendency for the population to equilibrate its genetic composition and resist sudden change is called genetic homoeostasis.

■ **genetic erosion**
1. the loss of genetic diversity within a population of the same species, the reduction of the genetic base of a species or the loss of an entire species over time.
2. the loss of genetic information that occurs when highly adapt cultivars are developed and threaten the survival of their more locally adapted ancestors, which form the genetic base of the crop.

■ **genetic fine structure**
the structure of the gene analysed at the level of the smallest units of recombination and mutation (nucleotides).

■ **genetic gain**
the change achieved by artificial selection in a specific trait. The gain is usually expressed as the change per generation or the change per year. It is influenced by selection intensity, parental variation and heritability.

■ **genetic homoeostasis**
See **genetic equilibrium.**

■ **genetic homology**
the identity or near identity of DNA sequences, genes or alleles.

■ **genetic information**
the information contained in a sequence of nucleotide bases in a nucleic acid molecule.

■ **genetic instability**
different mechanisms that give rise to phenotypic variation.

■ **genetic interaction**
the interaction between genes resulting in different phenotypic expressions.

■ **genetic linkage map**
a chromosome map showing the relative positions of the known genes on the chromosomes of a given species.

■ **genetic load**
1. the relative decrease in the mean fitness of a population due to the presence of genotypes that have less than the highest fitness.
2. the average number of lethal mutations per individual in a population.

▪ **genetic map**
a linear designation of sites within a chromosome or genome, based upon the various frequencies of recombination between genetic markers. See **linkage map**.

▪ **genetic mapping**
the process of determination of a genetic map.

▪ **genetic marker**
any phenotypic difference, controlled by genes, that can be used for studying recombination processes or selection of a more or less closely associated target gene.

▪ **genetic material**
all single- or double-stranded DNA carrying genetic information or that is a substantial part of the genetic information.

▪ **genetic nomenclature**
the designation of genes by abbreviated gene descriptions or symbols, usually a beginning capital letter of the abbreviation represents a dominant allele, while a small beginning letter refers to a recessive allele gene symbol.

▪ **genetic polymorphism**
an occurrence in a population of two or more genotypes in frequencies that cannot be accounted for by recurrent mutation.

▪ **genetic recombination**
a number of interacting processes that lead to new linkage relationships of genes.

▪ **genetic resistance**
resistance against pathogens or pests due to specific or general gene action.

▪ **genetic resources**
the gene pool in natural and cultivated stocks of organisms that are available for human exploitation.

▪ **genetic screening**
testing groups of individuals to identify defective genes capable of causing hereditary conditions.

▪ **genetic segregation**
segregation.

▪ **genetic sterility**
a type of male sterility conditioned by nuclear genes, as opposed to cytoplasmic sterility.

▪ **genetic stock**
a variety or strain known to carry specific genes, alleles or linkage groups.

▪ **genetic system**
the organisation of genetic material in a given species and its method of transmission from the parental generation to its filial generations.

- **genetic targeting**
See **gene targeting**.

- **genetic thinning**
in seed orchards, it refers to the removal of orchard genotypes based on their supposed breeding value rouging.

- **genetic variability**
individuals differing in their genotypes due to mutational, recombinational or selective mechanisms.

- **genetic variance (v_{gen})**
a portion of phenotypic variance that results from the varying genotypes of the individuals in a population. Together with the environmental variance, it adds up to the total phenotypic variance observed amongst individuals in a population. It is divided into additive (resulting from differences between homozygotes, V_{add}) and dominance variance (resulting from specific effects of various alleles in heterozygotes, V_{dom}). $V_{gen} = V_{add} + V_{dom}$, the quotient V_{add} / V_{dom} is termed heritability in a narrow sense.

- **genetic variation**
a phenotypic variance of a trait in a population attributed to genetic heterogeneity.

- **genetic vulnerability**
the susceptibility of genetically uniform crops to damage or destruction caused by outbreaks of a disease or pest or unusually poor weather conditions or climatic change.

- **Genetically Modified Organisms (GMO)**
a term, currently used most often in official discussions, that designates crops that carry new traits which have been inserted through advanced genetic engineering methods (e.g., flavour-saver tomato, roundup-ready soybeans, Bt cotton or Bt maize).

- **geneticist**
a specialist in genetics.

- **genic balance theory**
the theory of Bridges that the sex of a fruit fly is determined by the relative number of X chromosomes and autosomal sets.

- **geniculate**
bent abruptly like a knee.

- **genome**
all the genetic material in the *chromosomes* of a particular organism. Its size is generally given as its total number of *base pairs*.

- **genome**
the total genetic information carried by a single set of chro-

mosomes in a haploid nucleus, for example, genome sizes: lamda phage 48.5 kb, *Escherichia coli* 4,500 kb, yeast 1.6×10^4 kb, *Drosophila* 1.2×10^5 kb. In 2000, the total nucleotide sequence of a plant genome, *Arabidopsis thaliana,* was for the first time identified, the sequenced genome contains about 120 kb.

- **genome allopolyploids**
  See **allopolyploid.**

- **genome analysis**
  the study of the genome by combination of cytogenetics, karyotyping and crossing.

- **genome doubling**
  See **autopolyploidisation.**

- **genome formula**
  See **genome symbol.**

- **genome mutation**
  spontaneous or induced changes in the number of complete chromosomes that result either in polyploids or aneuploids.

- **genome projects**
  research and technology development efforts aimed at *mapping* and *sequencing* some or all of the *genome* of human beings and other organisms.

- **genome symbol**
  the description of specific genome by a symbol, usually a capital letter with or without a specification.

- **Genomic In Situ Hybridisation (GISH)**
  an in situ hybridisation technique that uses total genomic DNA of a given species as a probe and total genomic DNA of another species as a blocking DNA. It is based on fluorescence in situ hybridisation. It is a useful method to detect interspecific or intergeneric genome differentiation, chromosome rearrangements (translocations) and substitutions or additions.

- **genomic library**
  a type of DNA library in which the cloned DNA is from a genomic DNA of the plant and/or organism, since genome sizes are relatively large compared to individual cDNAs, a different set of vectors is usually employed in addition to plasmid and phage bacterial artificial chromosomes (BAC).

- **Genomic Selection (GS)**
  a method in which the number of polymorphic bands resembling the recurrent parent is used as the selection criterion.

- **genomics**
the scientific study of genes and their role in a structure, growth, health and disease of a plant (e.g., how a certain number of genes contributes to the shape, function and the development of the organism.)

- **genophore**
the chromosome (genetic material) of prokaryotes and viruses.

- **genotoxic**
it refers to substances and circumstances inducing mutants and damage of the heri material.

- **genotype**
the genetic constitution of an organism, as opposed to its physical appearance (phenotype). Usually, it refers to the specific allelic composition of a particular gene or set of genes in each cell of an organism, but it may also refer to the entire genome.

- **genotypic variance**
See **genetic variance (V_{gen})**.

- **genus**
a taxonomic grouping of similar species.

- **genus name**
name of a taxonomic group.

- **geometric mean**
the square root of the product of two numbers.

- **geophyte**
a land plant that survives an unfavourable period by means of underground food-storage organs (rhizomes, tubers, bulbs, e.g., onions, tulips, potato, asparagus, Jerusalem artichoke, etc.).

- **geotaxis**
oriented movement of a motile organism toward or away from a gravitational force.

- **germ**
the embryo or a collective name given to the embryonic roots and shoot and the scutellum tissue of the grain.

- **germ cell**
a sex cell or gamete (egg or spermatozoan). Haldane's law: the generalisation that if first generation hybrids are produced between two species, but one sex is absent, rare or sterile, that sex is the heterogamic sex.

germ line
the lineage of cells from which the gametes are derived and which, therefore, bridges the gaps between generations, unlike somatic cells in the body of an organism.

germ pore
an area or hollow in a spore wall through which a germ tube may come out.

germ tube
the filament that emerges when a spore germinates pollen tube.

germability
the degree of potential for germination germination test.

germicidal
it refers to any substance or condition that kills the embryo.

germinal
referring to the germ or germination.

germinal mutation
mutation occurring in cells that are destined to develop into gametes.

germinal selection
the selection during gametogenesis against induced mutations that retard the spread of mutant cells.

germination
the beginning of growth of a seed, spore or other structure, usually following a period of dormancy and generally in response to the return of favorable external conditions. When it takes place a root is produced that grows down into the soil, at the same time, a stem and leaves are growing upward.

germination test
a procedure to determine the proportion of seeds that are capable of germinating under particular conditions, commonly, a standard germination is conducted on a 400 seed sample at +25 °C for 7 days and seedlings are evaluated in accordance with the Association of Official Seed Analysts (AOSA, Rules for Testing Seeds). Several analysts evaluate each sample, all fungal species are identified and when abnormal seedlings exist the primary abnormal-type is noted germinator.

germinative cell
germ cell.

germinator
an apparatus with which seed germination is realised under more or less controlled conditions.

- **germ-line theory**
a theory to account for the high degree of antibody variability found in population. The germ-line theory suggests that every B lymphocyte has all the genes for every type of immunoglobulin but transcribes only one. See **somatic mutation theory**.

- **germplasm**
the hereditary material transmitted to offspring through the germ cells and giving rise in each individual to the cells.

- **germplasm bank**
See **gene bank.**

- **germplasm collection**
See **gene bank.**

- **Gibberellic Acid (GA)**
a group of growth-promoting substances, they regulate many growth responses and appear to be a universal component of seeds and plants short-straw mutant aleuron(e) layer.

- **gibberellin**
the generic name of a group of plant hormones that stimulate the growth of leaves and shoots, they tend to affect the whole plant and do not induce localised bending movements. They are thought to act either at a transcriptional level or as inducers of enzymes. First isolated from the fungus *Gibberella fujikuroi,* which causes the Bakanae disease in rice aleuron(e) layer short-straw mutant.

- **giemsa stain**
1. a complex of stains specific for the phosphate groups of DNA.
2. Banding.

- **gigantism**
abnormal overdevelopment due to an increase in cell size (hypertrophy), for example, in roots of crucifers infected with club root (*Plasmodiophora brassicae*) or abnormal overdevelopment as a result of an increase in the number of cells in response to a disease-production agent (hyperplasia), for example, witches broom, cankers, galls, leaf cure or scab.

- **ginned lint**
cotton fibres after they have been removed from the seed.

- **glabrous**
without hair or smooth.

- **gland**
organs or swellings that usually secrete a watery or characteristic substance, many oily and aromatic products are glandular in origin.

■ **glasshouse**
greenhouse.

■ **glassy grain**
hyaline grain.

■ **glaucous**
with waxy bloom present on the surface of the plant structure, a whitish, greyish or bluish appearance is often imparted.

■ **gliadin**
any prolamin and a simple protein of cereal grains that imparts elastic properties to flour. It is a monomeric molecule between 30,000 and 75,000 kDa, it is divided in alpha-, gamma- and omega-gliadins. It may form large polymeric structures as a result of intermolecular disulfide bonds.

■ **GLN**
glutamine (an amino acid).

■ **globulin**
one of a group of globular, simple proteins, which are insoluble or only sparingly soluble in water, but soluble in salt solutions. They occur in plant seeds (mainly in dicots), where they have a variety of functions (e.g., legumin, vignin, glycinin, vacilin or arachin).

■ **glomerule**
a very compact cyme, a cluster of flowers.

■ **glu**
glutamate (an amino acid).

■ **glucide**
carbohydrate.

■ **glucose**
dextrose.

■ **glucoside**
glucosides are soluble in water and alcohol, some of them are highly poisonous (e.g., saponin, from tung). They are found in vegetative organs and some in seeds (e.g., salicin in bark and leaves of willows, amygdalin in seeds of almonds, peaches or plums, sinigrin in black mustard, aesculin in horse chestnut seeds, quereitron in the bark of oaks.)

■ **glume surface**
the upper external surface of the broad wing of the glume (e.g., in wheat, which is described as being rough and/or smooth when scratched with a needle point.)

■ **glume(s)**
the outermost pair of bractlike structures of each spikelet, a chaff-like bract.

■ **glutamic acid**
an amino acid involved in purine biosynthesis, occasionally

added to plant tissue culture media, it may replace ammonium ions as the nitrogen source. It is of key importance in pollen growth in vitro.

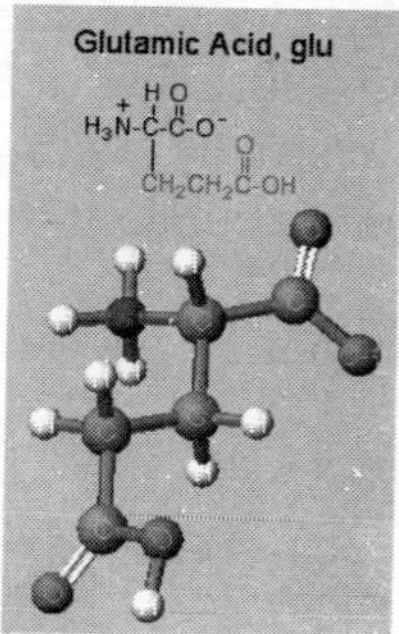

■ **glutathione**

a tripeptide containing glutamic acid, cystein and glycine capable of being alternately oxidised and reduced. It plays an important role in the cellular oxidation.

■ **gluten**

a term that is utilised to refer to a naturally occurring mixture of two different proteins (glutenin and gliadin) in the seeds of, for example, wheat. It is the principal protein in cereal seeds. It consists of a long polypeptide chain. In wheat, it possesses particular elasticity, which allows production of high-quality bread (strength and elasticity of the flour). More of the high-molecular-weight glutenin (which is 'stretchy' and imparts physical strength to a dough) results in a flour that is better suited to manufacture high-quality yeast-raised bread products.

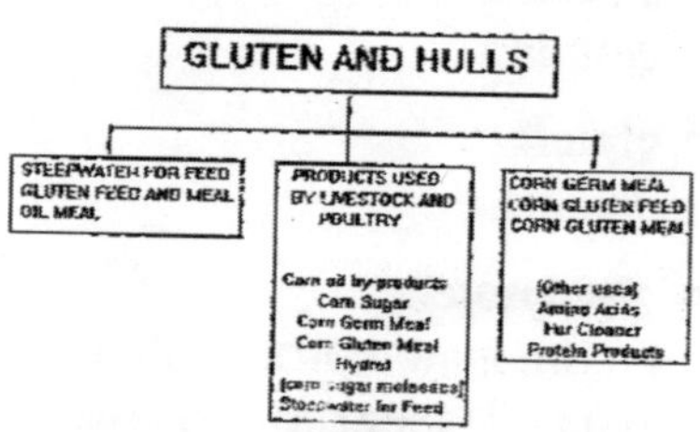

■ **glutenin**

it is soluble in aqueous or saline solutions or ethyl alcohol. It can also be extracted with strong acid or alkaline solutions. It is found in cereal seeds (e.g., glutenin in wheat or oryzinin in rice). It is divided in a low-molecular-weight glutenin subunit (LMW-GS) and in a high-molecular-weight glutenin subunit (HMW-GS) of about 65,000-90,000 kDa.

■ **gly**

glycine (an amino acid).

■ **glycerin(e)**

a three-carbon trihydroxy alcohol that combines with fatty acids to produce esters, which are fats and oils. It may serve as a cryoprotectant cryoprotectant refraction index.

■ **glycerol**

glycerin(e).

■ **glycine (gly)**
an amino acetic acid, the simplest alpha amino acid.

■ **glycinin**
globulin.

■ **glycoll**
glycine.

■ **glycoprotein**
a conjugated protein that consists of a carbohydrate covalently lined to a protein.

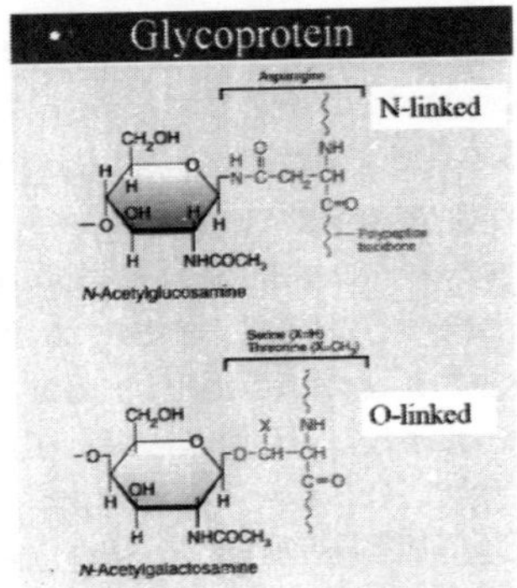

■ **glycoside**
glucoside.

■ **glycosylation**
the attachment of a carbohydrate to another molecule.

■ **glyoxaline**
imidazole.

■ **glyphosate**
a chemical compound used as a herbicide, glyphosate resistance is a subject of biotechnological approaches (e.g., tobacco and tomato transformants may show an over-expression of 5-enolpyruvyl shikimate-e-phosphate synthase [EPSPS], which is usually blocked at normal concentrations by the herbicide), transformed cells, callus or individuals proved to be tolerant to high glyphosate concentrations.

GLYPHOSATE (Roundup®)

$$OH{-}C({=}O){-}CH_2{-}NH{-}CH_2{-}P({=}O)(OH){-}OH$$

N-(phosphonomethyl)glycine
(isopropylamine salt

■ **GM food**
Genetically Modified Food.

■ **GMO**
Genetically Modified Organisms.

■ **GMO testing**
polymerase chain reaction detects the presence of a certain DNA sequence and is used to test maize, soybean, rice, cotton and canola tissue resulting in qualitative results. DNA is extracted from ground material, a certain section of the DNA is replicated numerous times to attain sufficient DNA to identify the promotor through electrophoresis electrophoresis sepa-

rates different size particles, allowing the 35S promoter to be detected when compared to a known 35S sample.

▪ **golgi apparatus**
a system of flattened, smooth-surfaced, membranaceous cisternae, arranged in parallel 20-30 nm apart and surrounded by numerous vesicles. A feature of almost all eukaryotic cells, this structure is involved in the packaging of many products of cell metabolism.

▪ **goodness of fit**
methods to test the conformity of an observed empirical distribution function of data with a posited theoretical distribution function (e.g., chi-square test) by comparing observed and expected frequency counts, the Kolmogorov-Smirnov test calculates the maximum vertical distance between the empirical and posited distribution functions.

▪ **gossypol**
a dark pigment, $C_{30}H_{30}O_8$, derived from cottonseed oil.

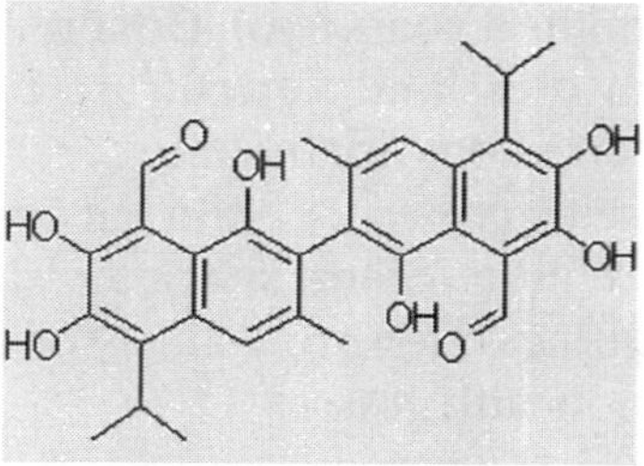

▪ **gradation**
the successive increase of organisms in a more or less cyclic or spontaneous pattern.

▪ **gradient**
a gradual change in some quantitative property over a specific distance.

▪ **graft**
to transfer a part (a small piece of tissue or organs) of an organism from its normal position to another position on the same organism (autograft) or to a different organism or species (heterograft). The stem or shoot that is inserted into a rooted plant is called the scion, the plant or part of a plant into which the scion is inserted is called the stock or understock. There are many different methods of grafting (e.g., flat grafting, split grafting or side grafting), sometimes the term is used in order to describe the point where a scion is inserted in the stock. See **transplant scion.**

▪ **graft chimera**
graft hybrid.

▪ **graft hybrid**
a plant made up of two genetically distinct tissues due to fusion of host and donor tissues after grafting.

- **grafting**
  the joining together of parts of plants, the united parts will continue their growth as one plant.

Grafting

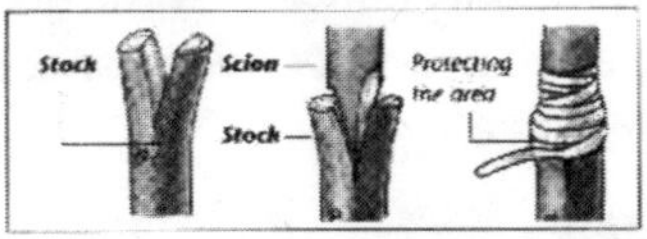

Stock Scion Stock Protecting the area

Budding

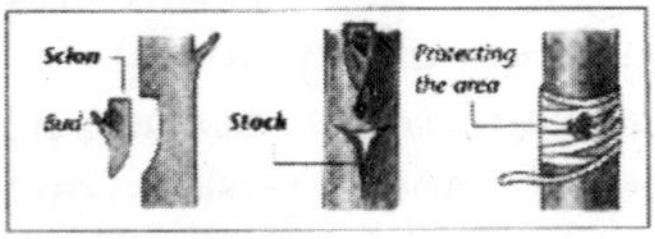

Scion Bud Stock Protecting the area

- **grafting tape**
  tape backed with biodegradable cloth, used in budding and grafting operations and in banding tree wounds.

- **grafting thread**
  a fine-waxed string used in budding and grafting operations.

- **grafting wax**
  a wax or related substance that is used to cover all injured parts of the rootstock and the scion after grafting and thus prevent infection by fungi or bacteria.

- **grain**
  a cereal caryopsis that may or may not be enclosed by the lemma an palea corn.

- **grain grade**
  a market standard established to describe the amount of contamination, grain damage, immaturity, test weight and marke traits.

- **Gram's stain**
  an important bacteriological staining procedure discovered empirically in 1884 by the Danish scientist C. GRAM, a technique used to distinguish between two major bacterial groups based on stain retention by their cell walls, bacteria are heat-fixed, stained with crystal violet, a basic dye, then with iodine solution, this is followed by an alcohol or acetone rinse, GRAM-positive bacteria are stained bright purple, GRAM-negative bacteria are decolorised, safranin is used to stain them.

- **gramineae**
  any of the monocotyledonous, mostly herbaceous plants, having jointed stems, slender sheathing leaves and flowers borne in spikelets of bracts the species in which the cereals are included appeared during the Cretaceous period (136-65 million years ago). Generally, in grassland agriculture the term does not include cereals when grown for grain but does include forage species of legumes often grown in association with grasses.

- **gramineous**
  gramineae.
- **granary**
  a storehouse or repository for grain.
- **granum**
  stacks of circular thylakoids, composed of lamellae in higher green plant chloroplasts, containing pigments and other essential components of photosynthetic light reactions.
- **grape sugar**
  dextrose.
- **grasses**
  gramineae.
- **gravitational water**
  the water that flows freely through the soil in response to gravity.
- **gravity separator**
  a machine utilising a vibrating porous deck and air flow to separate seed on the basis of their different specific gravities.
- **graze**
  the eating of crops by animals in the field.
- **green chop**
  green plants cut into small sections for animal feed.
- **Green Fluorescent Protein (GFP)**
  protein from the jelly fish, *Aequorea victoria* (Scyphozoa), the gene is used as a reporter gene. It fluoresces in UV light several variants have been developed, which all exhibit characteristic spectra, a significant advantages are that the protein can be seen in living tissue and are not toxic.
- **green manure crops**
  crops that are grown for the purpose of being ploughed into the soil to improve soil fertility and organic content. Phacelia may be a green manure crop in some regions, as are various legumes, such as lupins. The crops are ploughed into the soil while they are still green.
- **green manuring**
  green manure crops.
- **green revolution**
  advances in genetics, petroindustry and machinery that culminated in a dramatic increase in crop productivity during the third quarter of the twentieth century.
- **greenhouse**
  a building, room or area, usually of glass, in which the temperature is maintained within a

desired range, used for cultivating tender plants or growing plants out of season.

- **grey crescent**
  a cortical region of the newly fertilised egg of frogs and some salamanders that forms just after fertilisation on the side opposite sperm penetration.

- **grid design**
  for the grid design, plants or variants are divided into blocks and the best ones chosen from each.

- **grist**
  a mixture of grain (e.g., wheat or barley) utilised by, for example, the miller for grinding or the maltster for producing malt. Grist may contain a mixture of several varieties.

- **grits**
  See **semolina.**

- **groat**
  the caryopsis of oats after the husk has been removed.

- **ground germination rate**
  field germination.

- **ground state**
  the developmental state of a cell (or group of cells) in the absence of activation of a developmental regulatory switch.

- **group 1 intron**
  self-splicing intron that requires an external guanine-containing nucleotide for splicing, releases the intron in a linear form.

- **group coancestry**
  the probability that two genes taken from a gene pool of a population are identical by descent or the average of the cells in a coancestry matrix for the population concerned. See **coancestry.**

- **group ii intron**
  self-splicing intron that does not require an external nucleotide for splicing, releases the intron in a lariat form.

- **group merit**
  the genetic merit of a group as a function (weighted average or index) of its breeding value and gene diversity. It is an index to quantify the merit of a group as a weighted average of its advance in breeding value and its loss in gene diversity relative to some reference population.

- **group merit progress**
  group merit changes over generations, mainly as breeding produces a genetic gain but at the same time a loss, group merit progress (per year) takes genetic gain, gene diversity and

time into consideration and ought thus be a good measure of the progress in breeding.

- **group merit selection**
maximising the group merit of selections given the candidates and group merit measure. The selection method was initially called „population merit selection'.

- **group selection**
selection for traits that would be beneficial to a population at the expense of the individual possessing the trait.

- **group-selection method**
a method of regenerating and maintaining uneven-aged stands in which trees are removed in small groups.

- **growing tray**
a tray having compartments like an ice-cube tray, used for starting seeds.

- **growth analysis**
a mathematical analysis of crop or plant growth using relative growth rate, net assimilation rate, leaf area growth rate and crop growth rate.

- **growth curve**
a curve showing the change in the number of cells in a growing culture as a function of time.

- **growth form**
the form of a plant, the habit in which a plant grows (e.g., shrubby plant, climbing plant, leaf plant etc.)

- **growth habit**
the mode of growth of a plant, crops may be classified as (1) annuals (e.g., barley), (2) biennials (e.g., sugarbeet) or (3) perennials (e.g., alfalfa).

- **growth inhibitor**
any substance that retards the growth of a plant or plant part. Almost any substance will inhibit growth when concentrations are high enough. Common inhibitors are abscisic acid and ethylene. Other inhibitors, such as phenolics, quinones, terpens, fatty acids and amino acids affect plants at very low concentrations.

- **growth promoter**
a growth substance that stimulates cell division (e.g., cytokinin) or cell elongation (e.g., gibberellin).

- **growth rate (of crop)**
the crop growth rate is a specific plant growth analysis term denoting the absolute growth rate of mass per unit land, etc.

■ **growth regulator**

despite natural growth regulators, a synthetic compound that, when applied to a plant, promotes, inhibits or otherwise modifies the growth of that plant.

■ **growth stages syn developmental stage**

the discrete portion of the life cycle of a plant, such as vegetative growth, reproduction or senescence.

■ **growth substance**

a naturally occurring compound, other than a nutrient, that promotes, inhibits or otherwise modifies the growth of a plant.

■ **GS**

Genomic Selection.

■ **guanidine**

a crystalline, alkaline, water-soluble solid, CH_5N_3, used in making resins.

■ **guanine (g)**

a nitrogenous base, one member of the *base pair* G- C (guanine and *sine*).

nitrogen-containing base

phosphate group

Guanine (G)

five-carbon sugar

© 1997 Wadsworth Publishing Company/ITP

■ **guanosine**

the nucleoside having guanine as its base.

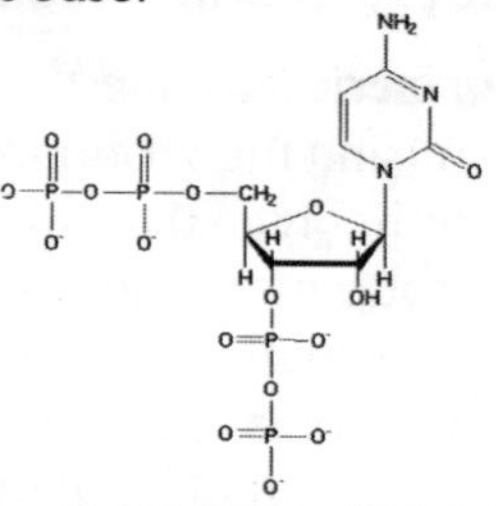

Guanosine 3',5'-tetraphosphate (ppGpp)

■ **guanylic acid**

guanosine monophosphate, a ribonucleotide constituent of ribonucleic acid that is the phosphoric acid ester of nucleoside guanosine.

■ **guard cell**

a specialised type of plant epidermal cells, two of which surround each stoma changes in their turgidity cause stomatal opening and closing.

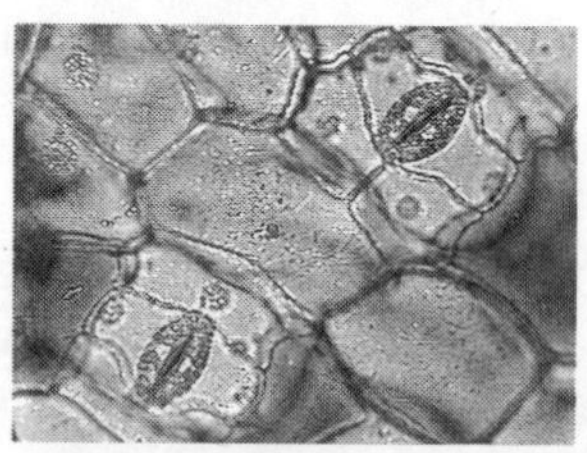

■ **guide RNA (gRNA)**

rNA that guides the insertion of uridines (RNA editing) into mRNAs in trypanosomes. Found in transcripts from minicircles and maxicircles of DNA in kinetoplasts.

- **guidepost**
  landmark.
- **GUS gene**
  a gene that codes for production of beta-glucuronidase (GUS protein) in *Escherichia coli* bacteria.
- **guttation**
  the tearlike extrusion of water and sometimes salts from the aerial parts of plants, particularly at night when transpiration rates are low, it is the process of water being exuded from hydathodes at the enlarged terminations of veins around the margins of the leaves in biotechnology, used for exudation of specific proteins made by artificially inserted genes.
- **gymnosperm(ous)**
  a kind of plant that produces seeds but not fruits. The seeds are not borne within an ovary and are called naked.
- **gynaecium**
  See **gynoecium**.
- **gynandromorph**
  an individual exhibiting both male and female sexual differentiation.
- **gynic**
  female.
- **gynodioecy**
  plant species or population in which female plants as well as hermaphroditic plants occur.
- **gynoecious**
  gynoecium.
- **gynoecium**
  the collective term for the female reproductive organs of a flower, comprising one or more carpels.

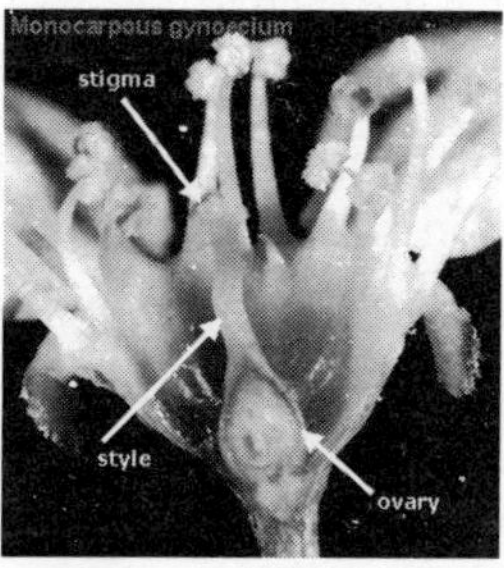

- **gynophore**
  the stalk that pushes pollinated peanut flowers into the soil.
- **habit(us)**
  the general appearance of a plant.
- **habitat**
  the living place of an organism or community characterised by its physical or biotic properties.
- **habituation**
  the diminishing requirement of some tissue cultures for growth-

regulatory substances, possibly due to endogenous production.

- **hair**
  a slender outgrowth of the epidermis, common on certain leaf or stalk structures.

- **hairy root disease**
  a disease in some dicots, root-like tissue is proliferated along segments of the stem. It is caused *Agrobacterium rhizogenes*, if it carries a Ri plasmid.

- **half-diallel cross**
  the crossing of a series of genotypes in all combination except reciprocal combinations.

- **half-grain method**
  a method in cereals separating the embryo from the endosperm by transversal dissection. It is applied when the endosperm is used for biochemical and molecular studies while the embryo can be grown for seed production. In this way preselected individuals and/or genotypes can be multiplied, it can reduce the breeding population and experimental costs.

- **half-hardy annuals**
  plants that will survive some frost, but not a long freeze.

- **half-life (time)**
  the time required for one-half the atoms of a given amount of a radioactive substance to decay.

- **half-shrub**
  suffrutex.

- **half-sib progeny selection**
  method of overstored seeds (*syn* remnant seed procedure).

- **halomorphic**
  a soil that has high levels of salt.

- **halophilic**
  See **halophilous.**

- **halophilous**
  a salt-loving organism, adapted to a high-salt environment (e.g., in a salt marsh.)

- **halophytes**
  plants that grow in saline soil.

- **hanahan transformation procedure**
  an optimised procedure for the transformation of *Escherichia coli* with plasmid DNA using $CaCl_2$.

- **hand weeding**
  manually removing the undesirable species inhibiting the growth of valued species.

- **hand-dibbed**
sowing individual seeds by hand according to a special plot design. It is mainly used for F1 seeds when they are rare or seeds that need special care.

- **hanging-drop culture (technique)**
a method of microscopic examination of organisms or particles suspended in a drop on a special concave microscopic slide.

- **haplodiploidy**
the sex-determining mechanism found in some insect groups among which males are haploid and females are diploid.

- **haploid**
a single set of *chromosomes* (half the full set of genetic material), present in the egg and sperm cells of animals and in the egg and pollen cells of plants. Human beings have 23 chromosomes in their reproductive cells. Compare *diploid*.

- **haploidisation**
production of a haploid from a diploid by progressive chromosome loss.

- **haplontic**
organisms in which meiosis occurs in the zygote resulting in four haploid cells.

- **haplophase**
that part of the life cycle in which the gametic chromosome number is found in reproduction cells.

- **haplosis**
the meiotic division resulting in haploid cells and/or gametes.

- **haplosomic**
the situation in which the homologue of a pair of chromosomes is missing in somatic cells monosomic.

- **haplotype**
a combination of alleles of closely linked loci found in a single chromosome. Sometimes, a combination of particular nucleotide variants within a given DNA sequence.

- **hard seed**
a seed that is dormant due to the nature of its seedcoat, which is impervious to either water or oxygen.

- **hard wheat**
tetraploid wheat, *Triticum durum*, used for high-quality noodles, bread and pastas.

- **hardening (off)**
the gradual process of acclimating plants started indoors to outside conditions (e.g., placing them in a sheltered location outdoors for increasing lengths

of time over a period of days), acclimatisation.

- **hardiness**
  the capability of a plant to withstand environmental stress.

- **hard-leafed**
  See **sclerophyllous.**

- **Hardy-Weinberg law**
  the concept that both gene frequencies and genotype frequencies will remain constant from generation to generation in an infinitely large, interbreeding population in which mating is at random and there is no selection, migration or mutation.

- **harlequin chromosome**
  Sister Chromatid Exchange (SCE).

- **harrow**
  an agricultural implement with spikelike teeth or upright disks, for levelling and breaking up clods in ploughed land or experimental plots, to draw a harrow over soil.

- **harvest index (HI)**
  the proportion of the biological yield to economic yield (i.e., grain.)

- **hat medium**
  a selection medium for hybrid cell lines, contains hypoxanthine, aminopterin, thymidine. Only cell lines expressing both hypoxanthine phosphoribosyl transferase (HPRT+) and thymidine kinase (TK+) can survive in this medium. Aminopterin inhibits de novo synthesis of nucleosides, while HPRT and TK supply them from hypoxanthine and thymidine.

- **haulm**
  stems or stalks collectively, as of peas, beans or hops or a single stem or stalk.

- **haustorium**
  in certain parasitic fungi, an outgrowth from hypha that penetrates a host cell in order to absorb nutrients from it, in some parasitic angiosperms, outgrowth of the roots, in endosperm development, nutrient-gathering outgrowths toward surrounding tissue of the developing endosperm.

- **hay**
  herbage, as grass, clover or alfalfa, cut and dried for use as forage.

- **head**
  an inflorescence in which the floral units on the peduncle are tightly clustered, surrounded by a group of flowerlike bracts called an involucre (e.g., in sunflower.)

- **head components**
generally, all components of the inflorescence of grain and grass crops.

- **head shattering**
a preharvest lost of kernels in cereals caused by loose seeds inside the spikelets and mechanical shattering (wind, etc.)

- **heading**
emerging spikes (i.e., from initial emergence of the inflorescence from the boot until the inflorescence is fully exerted) , in viticulture, to shorten or prune the trunk when it reaches the desired height, done in an effort to focus growth on the lower shoots.

- **heat filter**
absorption glass filter that attenuates infrared radiation, but transmits light in the visible wavelength range.

- **Heat Shock Protein (HSP)**
when certain plants are exposed to high temperature, heat shock proteins are synthesised. They provide thermal protection to subsequent heat stress stress protein(s).

- **heating**
gentle heating of the slide over a spirit or other flame flattens the cells, sticks them to the glass slide and cover slip and spreads the chromosomes, whether in prophase or metaphase.

- **heat-shock protein**
one of a number of proteins appearing in a cell after the cell has been subjected to elevated temperatures.

- **heaving**
lifting effect of the soil due to alternate freezing and thawing, it may result in the lifting up of plants and may tear them loose from the soil or may shear off roots.

- **heavy feeder**
a plant that requires a great amount of nitrogen because of its speedy growth (e.g., squash, potato, tomato, etc.).

- **heavy soil**
a soil that has a high content of clay and is difficult to cultivate.

- **hectare**
equals 10,000 square meters equals 2.471 acres.

- **hedgerow**
a row of bushes, trees or special plants forming a hedge.

- **heel**
a piece of the old branch or shoot that is detached from the old branch or shoot along with a cutting made.

■ **heeling in**

temporarily covering the base of a plant with soil for a short time (e.g., stecklings of sugarbeet during the winter) in order to prevent frost damage of cold-sensitive genotypes.

■ **heirloom plant**

a plant that was developed and in cultivation sometime in the past heirloom variety.

■ **heirloom variety**

heritage seeds.

■ **helicase**

a protein that unwinds DNA at replication forks.

■ **helix-turn-helix motif**

configuration, found in some DNA-binding proteins, consisting of a recognition helix and a stabilising helix separated by a short loop.

■ **hematoxylin staining**

a method used as an early indicator of aluminium toxicity effects on the apices of young developing roots of cereals grown in nutrient solution.

■ **hemialloploid**

not a normal full alloploid but a segmental alloploid, in which some parts of the unified genomes show some degree of structural conformity.

■ **hemiautoploid**

autopolyploids with a certain degree of differentiation between the diploid sets of chromosomes, either by a subsequent differentiation of previously homologous sets of chromosomes or by spontaneous or induced intervarietal or subspecies hybridisation.

■ **hemicellulose**

a heterogeneous group of compounds that in plant cell walls form part of the matrix within which cellulose fibers are embedded.

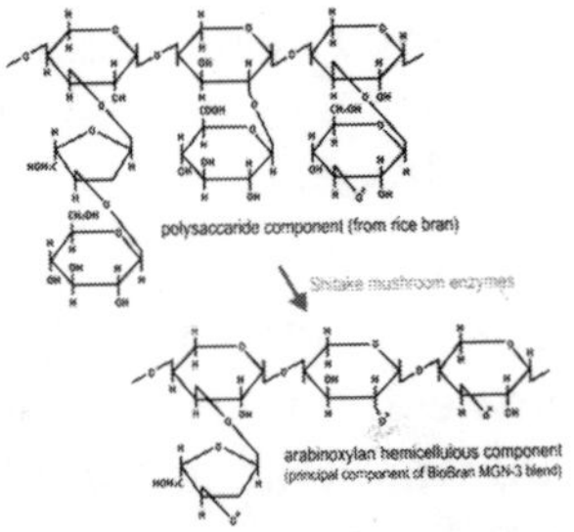

■ **hemichromosome**

a chromosome split into chromatids without previous reduplication at interphase.

■ **hemigamy**

where gametes fuse but the nuclei do not, forming a dikaryon.

■ **hemihaploid**

individuals with half of the normal haploid chromosome number.

- **hemimethylation**
  the state in which a DNA duplex is methylated in one strand but unmethylated in the other.

- **hemiploid**
  individuals with half of the somatic chromosome number haploid.

- **hemizygous**
  the condition of loci on the X chromosome of the heterogametic sex of a diploid species or more generally when one part of the genome, in a normally diploid species, is present in only one copy.

- **herb(s)**
  a small, nonwoody, seed-bearing plant in which all the aerial parts die back at the end of each growing season.

- **herbaceous**
  nonwoody, as applied to kinds of plant growth herb(s).

- **herbage**
  plant material used for animal feed.

- **herbarium**
  a collection of dried, preserved and systematically classified plants.

- **herbarium beetle**
  cartodere filum, eats the spores of certain fungi (e.g., *Lycoperdon*, smuts, etc.) that are attached to the plant material of a herbarium.

- **herbarium glue**
  an adhesive that minimises cracking, discoloration and shattering with age. It is used in fastening plant specimens to the herbarium sheet.

- **herbarium paste**
  herbarium glue.

- **herbicide**
  a chemical substance that suppresses or eliminates plant growth, it may be a nonselective or selective weed killer.

- **herbicide tolerance**
  the ability of some plants to tolerate herbicides. It is a task of genetic engineering to modify crop plants for this trait in order to apply herbicides against weeds in the field.

- **hercogamy**
  See **herkogamy**.

- **hereditary**
  transmissible from parent to offspring or progeny.

- **hereditary determinant**
  any genetically acting unit of an organism that is replicated

and conserved, transferred from generation to generation.

- **hereditary factor**
See **hereditary determinant.**

- **heredity**
the transmission of genetic characters from one generation to the next generation, it operates primarily by the germ cells in sexually reproducing species.

- **heri**
hereditary.

- **heritability**
a measure of the degree to which a phenotype is genetically influenced and can be modified by selection. It is represented by the symbol h^2, this equals V_{gen}/V_{phe} where V_{gen} is the variance due to genes with additive effects and V_{phe} is the phenotypic variance there are two types of heritability: (1) broad-sense heritability, $h_b b^2 = V_{gen}/V_{phe}$ and (2) narrow-sense heritability, $h_n b^2 = V_{add}/V_{dom}$.

- **heritability in the narrow sense**
the proportion of phenotypic variance that can be attributed to additive genetic variance.

- **heritage seeds**
nonhybrid seeds of old varieties that have been passed from generation to generation.

- **herkogamy**
pollination by the neighbour individual, population or species.

- **hermaphrodite**
a plant having both female and male reproductive organs in the same flower.

- **hermaphroditic**
reproductive organs of both sexes present in the same individual or in the same flower in higher plants bisexual.

- **hesperidia**
a berrylike fruit with papery internal separations or septa and a leathery, separable rind (e.g., orange, lemon, lime, grapefruit).

- **heteroallele**
an allele that differs from other alleles of the same gene by nucleotide differences at different sites within the gene, in contrast with 'true' alleles, of which only four are possible at each site within the gene.

- **heteroauxin**
an obsolete term for the auxin 1H-indole-3-acetic acid (IAA) indole-3-acetic acid.

- **heterochromatic**
of chromosome regions or whole chromosomes that have a dense, compact structure in telophase, interphase and early prophase.

heterochromatin
the chromosome material that accepts stains in the interphase nucleus (unlike euchromatin). Such regions, particularly those containing the centromeric and nucleolus organisers, may adhere to form a chromocenter. Some chromosomes are composed primarily of heterochromatin, these are termed heterochromosomes, such as Y chromosome in some species.

heterochromosome
any chromosome that differs from the autosomes in size, shape and behaviour.

heteroduplex
a DNA double helix formed by annealing single strands from different sources. If there is a sequence difference between the strands, the heteroduplex may show single strand loops or bubbles (unpaired regions.)

heteroduplex analysis
duplex DNA formed by strands from different sources, referred to as a heteroduplex, will have loops and bubbles in regions where the two DNAs differ. Electron microscopic observation (analysis) of this DNA has been a useful tool in recombinant DNA work.

heteroduplex DNA
a double-stranded DNA molecule formed by the annealing of strands from two different sources, as opposed to homoduplex, which has homologous strands. As a result, there are regions noncomplementary and showing abnormalities in the form of extra loops.

heteroduplex DNA model
a model that explains both crossing over and gene conversion by assuming the production of a short stretch of heteroduplex DNA (formed from both parental DNAs) in the vicinity of a chiasma.

heteroduplex mapping
the use of heteroduplex analysis to determine the location of various insertions, deletions or heterogeneities between two DNA molecules.

heteroecious
1. a species that produces male and female gametes on different individuals dioecious.
2. the requirement of a pathogen for two host species to complete its life cycle.

heterofertilisation
fertilising of the nuclei of endosperm and embryo-forming

cells by genetically different gametes.

- **heterogam(et)ic**
a species that sexually reproduces by two types of gametes.

- **heterogamete**
anisogamete.

- **heterogametic sex**
the sex with (usually) two heteromorphic (differently shaped) sex chromosomes (for example, X and Y). During meiosis it produces two kinds of gametes in regard to these sex chromosomes.

- **heterogamy**
reproduction involving two types of gametes.

- **heterogeneity**
the production of identical or similar phenotypes by different genetic mechanisms.

- **heterogeneity index**
a measure for genetic differences within populations.

- **heterogeneous nuclear mrna (hnrna)**
the original RNA transcripts found in eukaryotic nuclei before post-transcriptional modifications. A diverse assortment of RNA types found in the nucleus, including mRNA precursors (pre-mRNA) and other types of RNA.

- **heterogenetic**
chromosome pairing between more or less different genomes in amphiploids.

- **heterogenic**
gametes or populations differing in alleles or genes.

- **heterograft**
heterologous graft, the scion and rootstock derive from different species.

- **heterohistont**
an individual or cell aggregate that is composed of tissues of genetically different origin.

- **heterokaryon**
a cell that contains two or more nuclei from different origins. A cell composed of two different nuclear types in a common cytoplasm.

- **heterokaryon test**
a test for cytoplasmic organelle mutations (e.g., mitochondrial or chloroplast mutations), based on new associations (recombination) of phenotypes in cells derived from genetically marked haploid heterokaryons. Since the heterokaryons are haploid and produce haploid spores by a process that does not involve diploid nuclei, normal genetic recombination does not occur. Therefore evi-

dence of recombination is evidence of re-association by some other means such as re-association of an organelle with nucleus.

- **heterokaryotype**
  a chromosome complement that is heterozygous for any sort of chromosome mutations.

- **heterolabelling**
  a chromosomal labelling pattern due to induced or spontaneous exchange of labelled and nonlabelled half-chromatids.

- **heterologous gene expression**
  expression of a gene in another host.

- **heterologous probing**
  probing at low stringency with a DNA fragment that originates from another organism and thus does not have an identical counterpart in the target DNA. Often it gives significant signals because of sequence conservation.

- **heteromeric**
  genes that control the determination of a trait by joint gene action, but each of them shows a definitely different contribution to the final product.

- **heteromorphic**
  chromosomes that differ in size and/or shape, the term is also used for meiotic pairing configuration, which is composed of different chromosomes and/or chromosome segments.

- **heteromorphic bivalent**
  a bivalent consisting of nonhomologous chromosomes or segments.

- **heteromorphic chromosomes**
  a chromosome pair with some homology but differing in size, shape or staining properties. Homologous chromosome pair which are not morphologically identical (e.g., the sex chromosomes).

- **heteromorphous**
  See **heteromorphic.**

- **heteromorphy**
  See **heteromorphic.**

- **heterophylly**
  the production of more than one leaf form on a plant species, in developmental heterophylly, juvenile leaves may differ from adult ones.

- **heterophyte**
  a plant that is dependent upon another, obtaining its nourishment from other living or dead organisms, such as parasites or saprophytes.

- **heteroplasmic**
  alloplasmic.

- **heteroplasmon**
  1. a cell containing a mixture of genetically different cytoplasms, generally different mitochondria or different chloroplasts.
  2. different mitochondria or different chloroplasts.

- **heteroplasmonic**
  alloplasmic.

- **heteroplasmy**
  the existence within an organism of genetic heterogeneity within the populations of mitochondria or chloroplasts.

- **heteroplastidic**
  cells whose plastids are different in shape.

- **heteroploid**
  deviating chromosome numbers from the standard chromosome set.

- **heteropycnotic**
  chromosomes or chromosomal segments that show a different coiling or staining pattern.

- **heterosis**
  the increased vigour of growth, survival and fertility of hybrids, as composed with the two homozygotes. It usually results from crosses between two genetically different, highly inbred lines, it is always associated with increased heterozygosity. In breeding, three types of heterosis are distinguished: (1) F1 yielding more than the mean of the parents, (2) F1 yielding more than the best yielding parents, (3) F1 yielding more than the best yielding variety, for the genetic basis of heterosis two hypotheses have received the most attention: dominance hypothesis and overdominance hypothesis.

- **heterosomal**
  a chromosome mutation involving non-homologous chromosomes.

- **heterosome**
  a chromosome that is deviating from the standard chromosomes in size, shape or behaviour.

- **heterospory**
  the production of spores of two different types of the same plant.

- **heterostyly**
  a polymorphism among flowers that ensures cross-fertilisation through pollination by visiting insects, flowers have anthers and styles of different length.

- **heterothallic**
  a botanical term used for organisms in which the two sexes reside in different individuals.

- **heterothallic fungus**
  a fungus species in which two different mating types must unite to complete the sexual cycle.

- **heterotroph**
  organism requiring an organic form of carbon as a carbon source.

- **heterozygosity**
  a measure of the genetic variation in a population, with respect to one locus, stated as the frequency of heterozygotes for that locus.

- **heterozygote**
  an individual having a heterozygous gene pair. A diploid or polyploid with different alleles at a particular locus.

- **heterozygote advantage**
  a selection model in which heterozygotes have the highest fitness.

- **heterozygotic**
  heterozygous.

- **heterozygous**
  the condition of having unlike alleles at corresponding loci.

- **heterozygous gene pair**
  a gene pair having different alleles in the two chromosome sets of the diploid individual for example, Aa or, A1A2.

- **hexaploid**
  a cell having six chromosome sets or an organism composed of such cells.

- **hexasomic**
  a cell or individual showing one chromosome six times.

- **hexokinase**
  an enzyme that catalyses the phosphorylation of hexose sugars.

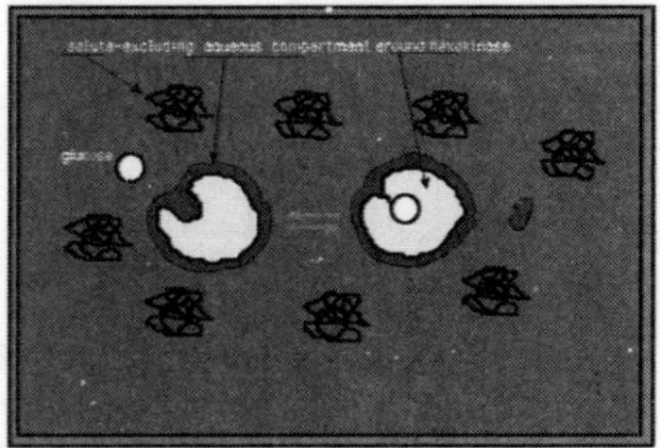

- **hexose**
  a monosaccharide sugar that contains six carbon atoms.

- **hibernaculum**
  the winter resting body of some plants, generally a bud-like arrangement of potential leaves.

- **hieracium type**
  apospory where the embryo sac is derived from a somatic cell, usually from the centre of the nucellus, three mitoses lead to a mature eight-nucleate, unreduced embryo sac.

- **hierarchical classification**
  the grouping of individuals by a series of subdivisions or ag-

glomerations to form characteristic 'family trees' or dendrogram of group.

■ **high oleic**
in sunflower breeding, seeds that contain a trait for high oleic fatty acid content in its oil, a premium oil used in the snack food industry.

■ **high-velocity microprojectile transformation**
a procedure for introduction of DNA into plant cells, for example, gold particles are coated with DNA and propelled at high speed through the target cell walls by means of an electrical or gunpowder discharge.

■ **hill plot**
the scar on a seed that marks the point at which it was attached to the plant.

■ **Himalayan**
a mammalian temperature-dependent coat phenotype, generally albino with pigment only at the cooler tips of the ears, feet and tail.

■ **hips**
seed pods (fruits), for example, in roses or apples, that are formed after a flower's petals fall if the bloom was pollinated.

■ **his**
histidine (an amino acid).

■ **histocompatibility antigen**
antigen that determines acceptance or rejection of a tissue graft by the immune system.

■ **histocompatibility gene**
a gene encoding an histocompatibility antigen.

■ **histogram**
a bar graph of a frequency distribution in which the bars are displayed proportionate to the corresponding frequencies.

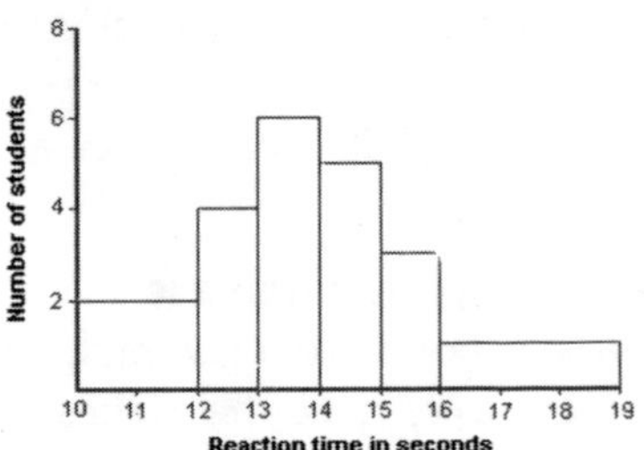

■ **histone**
a type of basic protein that forms the unit around which DNA is coiled in the nucleosomes of eukaryotic chromosomes. Arginine and lysine rich basic proteins making up a substantial portion of eukaryotic nucleoprotein.

■ **hitchhiking effect**
genes favoured in a population by close linkage with other

genes, which are positively selected.

- **holandric trait**
trait controlled by a locus found only on the Y chromosome. Involves father to son transmission.

- **hole seeding**
dibbling (seed).

- **holocentric**
applied to chromosomes with diffuse centromeres such that the properties of the centromere are distributed over the entire chromosome.

- **holoenzyme**
the complete enzyme including all subunits. Often used in reference to RNA and DNA polymerases.

- **holokinetic**
holocentric.

- **homeo-box**
a consensus sequence of about 180 base pairs discovered in homeotic genes in *Drosophila*. Also found in other developmentally important genes from yeast to human beings. A family of quite similar 180 base-pair DNA sequences that encode a polypeptide sequence called a homeo-domain, a sequence-specific DNA binding sequence. While the homeobox was first discovered in all homeotic genes, it is now known to encode a much more widespread DNA-binding motif.

- **homeo-domain**
an approximately sixty-amino acid protein domain translated from the homeo-box. A highly conserved family of protein domain sequences 60 amino acids in length found within a large number of transcription factors that can form a helix-turn-helix structure and bind DNA in a sequence-specific manner.

- **homeologous chromosomes**
partially homologous chromosomes, usually indicating some original ancestral homology.

- **homeosis**
the replacement of one body part by another. Homeosis can be caused by environmental factors leading to developmental anomalies or by mutation.

- **homeostatic mechanism**
a physiological process that contributes to the maintenance of a relatively s internal environment in a multicellular organism.

- **homeotic gene(s)**
a class of genes that determines the identity of an organ, segment or other structural unit during development. It controls

the identity of, for example, floral organs.

- **homeotic mutation**
  a mutation that causes one body structure to be replaced by a different body structure during development.

- **homoallelic**
  applied to allelic mutants of a gene that have different mutations at the same site.

- **homoeologous group**
  series of two or more chromosomes with similar but not homologous chromosomes (e.g., in hexaploid wheat or oat.)

- **homoeostasis**
  the tendency of a biological system to resist change and to maintain itself in a state of s equilibrium.

- **homogametic**
  producing only male or female gametes.

- **homogametic sex**
  the sex with homomorphic (similarly shaped) sex chromosomes (for example, XX), producing only one kind of gamete in regard to the sex chromosomes.

- **homogamic**
  of matings between individuals from the same population or species.

- **homogamous**
  hermaphroditic flowers showing synchronised function of male and female sex organs.

- **homogamy**
  the preference of individuals to mate with others of a similar genotype or phenotype, in botany, the condition in which male and female parts of the flower mature simultaneously.

- **homogenic**
  in cytology, sometimes it refers to chromosome pairing between morphologically identical genomes in amphiploids.

- **homologies**
  similarities in DNA or protein sequences between individuals of the same species or among different species.

- **homologous**
  applied to organs and chromosomes, both showing identical structures.

- **homologous chromosomes**
  a pair of chromosomes containing the same linear sequences, each derived from one parent.

- **homologous genes**
  genes with a common ancestor, generally used to describe genes from different species but which are similar and have the same function.

- **homologous recombination**
breakage and reunion between homologous lengths of DNA mediated in *Escherichia coli* by products of the genes RecA and RecBCD and functionally equivalent genes in other organisms.

- **homologue**
a member of a pair of homologous chromosomes.

- **homomeric**
genes that control the determination of a trait by joint gene action and each of them shows a similar contribution to the final product.

- **homomorphic bivalent**
a bivalent that is composed by two homologous chromosomes (i.e., in size and shape).

- **homomorphic chromosomes**
morphologically identical members of a homologous pair of chromosomes.

- **homoplasmic**
cells or individuals that carry two or more different types of cytoplasmic components.

- **homoplasmonic**
See **homoplasmic**.

- **homoplasmy**
the existence within an organism of only one type of plastid, usually referring to genetic identity of mitochondria or chloroplasts.

- **homopolymer-tailing**
attachment of identical nucleotides to the 3' end of a DNA molecule that can be achieved with terminal deoxynucleotide transferase.

- **homospory**
a condition in which an organism produces only one type and size of spore, e.g., microspores.

- **homothallic**
a botanical term used for groups whose individuals are not of different sexes.

- **homothallic fungus**
a fungus species in which a single sexual spore can complete the entire sexual cycle. (compare **heterothallic fungus**).

- **homothallic strains**
in yeast, strains that are capable of switching mating type and thus of mating with themselves, forming zygotes, asci and spores under appropriate conditions.

- **homozygosity**
the presence of identical alleles at one or more loci in homologous chromosomal segments.

■ **homozygote**
an individual having a homozygous gene pair. A diploid or a polyploid with identical alleles at a locus.

■ **homozygous**
having identical rather than different alleles in the corresponding loci of a pair of chromosomes and therefore breeding true.

■ **homozygous gene pair**
a diploid gene pair having identical alleles in both copies for example, AA or, aa.

■ **honeybee**
any bee that collects and stores honey (e.g., *Apis mellifera*) and contributes to improved seed-setting in cross-pollinating crops (e.g., rapeseed, fruit trees, etc.)

■ **honeycomb design**
in a honeycomb design, the plant at the center of the hexagon is compared with every other plant within the hexagon, a plant is chosen only if it is superior to every other plant in the hexagon. It was developed for selecting individual plants in a population. Seeds or plants are usually spaced equidistantly from one another in a hexagon pattern, plants are spaced far enough apart that they cannot compete with adjacent individuals, homogeneous checks can be included. The size of the hexagon determines the selection intensity, it is used to minimise adverse effects of interplant competition.

■ **honeydew**
a sticky exudate containing conidia, which is produced during one stage of the life cycle of the fungus *Claviceps purpurea* ergot.

■ **hook climber**
a plant that climbs by the aid of hooks or prickles (e.g., roses) climbing plants.

■ **hordecale**
an amphiploid hybrid between cereal species of the genera *Hordeum* and *Secale,* in which *Hordeum* species served as donors of the cytoplasm.

■ **hordein**
prolamin.

■ **hordeum bulbosum procedure**
a method for producing zygotic haploids in barley by crossing *Hordeum bulbosum* with *Hordeum vulgare* genotypes. After formation of zygotes the *H. bulbosum* chromosomes are subsequently eliminated during embryogenesis. The chromosomes of the wild barley are subsequently eliminated, which results in haploid *H. vulgare* plants.

- **horizontal resistance**
  resistance conditioned by polygenes or quantitative genes, it is race nonspecific in nature and does not reveal a gene-for-gene hypothesis, the type of resistance is difficult to identify.

- **horticulture**
  the science or art of cultivating flowers, fruits, vegetables or ornamental plants (e.g., in a garden, orchard or nursery.)

- **hortus siccus**
  a collection of specimens of plants carefully dried, preserved and described for botanical purposes and comparisons of mutants, etc. herbarium.

- **host**
  a living organism harbouring a parasite and its cells and metabolism are used for the growth of pathogens. Plant host can be classified by: (1) importance: (a) primary or principal hosts, (b) secondary hosts, (c) intermediate hosts, (d) accessory hosts, (e) accidental hosts and (f) definite or final hosts, (2) season: (a) winter hosts and (b) spring hosts, (3) other functions: (a) alternative, alternate or differential hosts, (b) transport or transfer hosts.

- **host cell**
  a cell whose metabolism is used for the growth of a pathogen.

- **host plant**
  See **host**.

- **host range**
  the spectrum of genotypes that can infect a specific pathogen or pest. In molecular biology, hosts in which a phage or plasmid can replicate, restriction is one factor that can limit the host range of plasmids or phages.

- **host species**
  See **heteroecious**.

- **host-mediated restriction**
  a mechanism by which bacteria prevent infection by phages originating from other bacteria, restriction also acts against unmodified plasmid DNA in transformation experiments, restriction endonucleases cleave the foreign DNA while the host DNA is protected from cleavage by specific methylation.

- **host-parasite specificity**
  the ability of a pathogen to pathogenise a specific group of plants.

- **hot spot**
  in genetics, one of sites tending to mutate frequently.

- **housekeeping enzymes**
enzymes present in all cells capable of normal metabolism, they are essential for the synthesis or breakdown of proteins, nucleic acids and lipids, for glycolysis and respiration and for many standard metabolic pathways.

- **housekeeping genes**
genes whose products are required by all cells at all times.

- **hull**
usually the hard, tightly adhering, outer covering of a seed or caryopsis, which is composed of the pericarp in some species and the lemma and palea in others , the persistent calyx at the base of some fruits, such as strawberry.

- **huller-scarifier**
a seed conditioning machine, it removes hulls or pods from seeds by an abrading or rubbing action.

- **humification**
the development of humus from dead organic material.

- **humus**
decomposed organic matter of soils.

- **husk**
the leaf sheaths of an ear of maize, the lemma and palea in other grass species or the dry outer cover of a coconut, the dry outer covering of some fruits or seeds (e.g., cereal grain.)

- **H-Y antigen**
the Histocompatibility Y-antigen, a protein found on the cell surfaces of male mammals.

- **hyaline**
clear, transparent.

- **hyaline grain**
hyaline.

- **hybrid**
any sort of sexual or somatic combination of genetically more or less differentiated parental cells, individuals or taxa. Specifically, an individual plant from a cross between parents of differing genotypes, any heterozygote represents dissimilar alleles at a given locus or a hybrid graft.

- **hybrid (general)**
the first-generation progeny of a cross between two different parents. An intermediate plant resulting from the crossing of two or more different bioytypes of the same species or biotypes from two different species.

- **hybrid breakdown**
hybrid lethality.

- **hybrid breeding**
  the discovery of heterosis has been recognised as one of the major landmarks of plant breeding. In comparison to inbred lines and homozygous material, the phenotypic superiority of heterozygotes is the basis of hybrid breeding. It is exploited for production of hybrid, synthetic and composite varieties. Hybrid breeding can be performed using traditional breeding techniques or the process can be hastened using gene marker technology to rapidly identify parents with desired genes for certain attributes. Numerous commercial crops are hybrids with increasing tendency. Seeds from a hybrid variety, if planted, will not deliver the same benefits as the original seeds and after several offspring will have lost the desired qualities from the original hybridisation.

- **hybrid chlorosis**
  plant and/or leaf chlorosis due to interacting genes and/or cytoplasm of parental lines in a hybrid.

- **hybrid complex**
  masking morphological differences of parental lines in a hybrid plant.

- **hybrid DNA**
  DNA whose two strands have different origins.

- **hybrid dysgenesis**
  a syndrome of effects including sterility, mutation, chromosome breakage and male recombination in the hybrid progeny of crosses between certain laboratory and natural isolates of *Drosophila*.

- **hybrid inviability**
  reduced vigour of hybrid plants compared to their crossing parents.

- **hybrid lethality syn hybrid sterility**
  the failure of hybrids to produce viable offspring.

- **hybrid necrosis**
  hybrid lethality.

- **hybrid plant**
  See **hybrid**.

- **hybrid plasmid**
  a plasmid that contains an inserted piece of foreign DNA.

- **hybrid seed production**
  the production of hybrid seeds by combination of more or less defined parental forms, usually those hybrid seeds are more productive or more sui than pure lines. They are used for subsequent growing and commercial production of a crop.

- **hybrid selection**
  the process of choosing plants possessing desired traits among a hybrid population.

- **hybrid sterility**
  the failure of hybrids to produce viable offspring.

- **hybrid variety**
  a variety produced from the cross fertilisation of inbred lines with favourable combining ability, the progeny is homogeneous and highly heterozygous, it can be produced by (a) two inbred lines, (b) single crosses, (c) a single cross and an open-pollinated or a synthetic variety or (d) two selected clones, seed lines, varieties or species.

- **hybrid vector/hybrid vehicle**
  an episome or plasmid containing an inserted piece of foreign DNA.

- **hybrid vigour**
  the increase in vigour of hybrids over their parental inbred types heterosis.

- **hybrid weakness**
  the decrease in vigour of hybrids below their parental inbred types, e.g., in rice a hybrid weakness phenomenon is controlled by a set of complementary genes, *Hwc1* (hybrid weakness c) and *Hwc2*. The *Hwc2* gene is prevalent among temperate Japonica rice but not among tropical Japonica or Indica rices, the chromosomal location of the *Hwc2* locus was determined from the segregation in the F1 hybrids made between recombinant inbred lines and the cultivar 'Jamaica' *Hwc2* was located between the two restriction fragment length polymorphism loci, *XNpb264* and *XNpb197* on chromosome 4.

- **hybrid zone**
  geographical region in which previously isolated populations that have evolved differences come into contact and form hybrids.

- **hybridisation**
  the process of joining two complementary strands of DNA or one each of DNA and RNA to form a double- stranded molecule.

- **hybridisation in situ**
  finding the location of a gene or gene product by adding specific radioactive or chemically tagged probes for the gene and detecting the location of the radioactivity or chemical on the chromosome or in the cell after hybridisation.

■ **hybridisation probe**
labelled nucleic acid molecule used to detect complementary DNA sequences after hybridisation.

■ **hybridise**
1. to form a hybrid by performing a cross.
2. to anneal nucleic acid strands from different sources.

■ **hybridoma**
a cell resulting from the fusion of a spleen cell and myeloma cell. These cells can be cloned and maintained indefinitely in cell culture and produce monoclonal antibodies.

■ **hydathode**
an epidermal structure specialised for the secretion or exudation of water guttation.

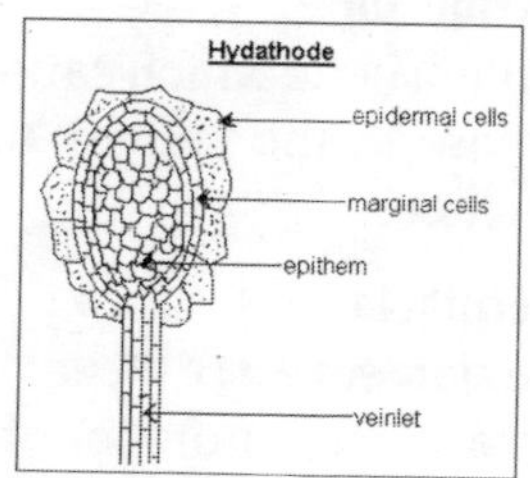

■ **hydratation**
the status of imbibition of the cytoplasm.

■ **hydrate**
any of a class of compounds containing chemically combined water.

■ **hydration**
the process whereby a substance takes up water.

■ **hydraulic seeding**
a method of planting grass seed by spraying it in a stream of water, which may contain other materials such as nutrients.

■ **hydrogen bond**
a weak bond involving the sharing of an electron with a hydrogen atom, hydrogen bonds are important in the specificity of base pairing in nucleic acids and in the determination of protein shape.

■ **hydrogen peroxide test**
a quick test to determine seed viability, in response to a hydrogen peroxide soak, viable seeds elongate their roots through a cut in the seedcoat, frequently used in conifer seeds.

■ **hydrolase**
an enzyme that catalyses reactions involving the hydrolysis of a substrate.

■ **hydrolysat(e)**
any compound formed by hydrolysis.

■ **hydrolysis**
in soil science, the process whereby hydrogen ions from water are exchanged for cat-

ions such as sodium, potassium, calcium and magnesium and the hydroxyl ions combine with the cations to give hydroxides.

- **hydrophytes**
  See **hygrophytes.**

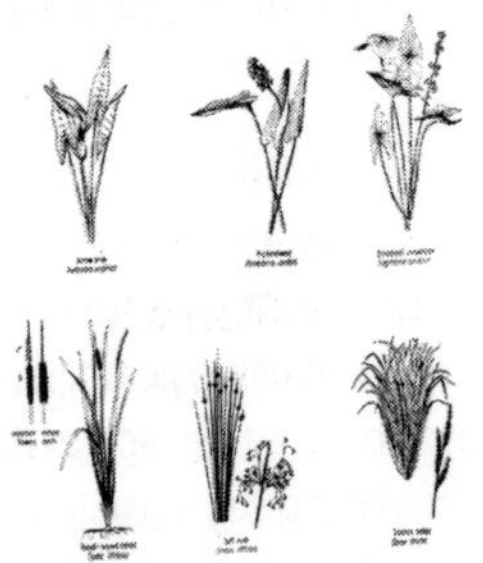

- **hydroponics**
  the cultivation of plants by placing the roots in liquid nutrient solutions rather than in soil.

- **hydroseeding**
  dissemination of seed hydraulically in a water medium, supplemented by mulch, lime and fertiliser.

- **hydrotaxis**
  movement of an organism toward or away from water.

- **hydroxide**
  a chemical compound containing the hydroxyl group.

- **hydroxyapatite**
  a form of calcium phosphate that binds double-stranded DNA.

- **hygrometer**
  any instrument for measuring the water-vapor content of the atmosphere.

- **hygrophytes**
  plants that can tolerate excess of water.

- **hygroscopic water**
  water that is adsorbed on to a surface from the atmosphere.

- **hygroscopic(al)**
  water attracting, plants or part of them becoming soft in wet air and hard in dry air.

- **hypanthium**
  a cup-like or tube-like enlargement of the floral receptacle or base of the perianth that surrounds the gynoecium and fruits.

- **hypermorph**
  a mutant gene which causes an increase in the activity that it influences.

- **hyperplasia**
  the enlargement of tissues by an increase in the number of cells by cell division.

- **hyperploid**
  aneuploid containing a small number of extra chromosomes.

- **hyperploidy**
  having additional chromosome complements com-

pared to the standard chromosome set.

- **hypersensitive resistance**
See **hypersensitivity.**

- **hypersensitive site**
a region of DNA located in a chromatin structure that makes it more sensitive to attack by endonucleases than DNA sites, located elsewhere in the chromatin. The presence of hypersensitive sites is correlated with transcription of adjacent DNA sequences in eukaryotic cells.

- **hypersensitivity**
resistance.

- **hypersensitivity**
the response to attack by a pathogen of certain host plants in which the invaded cells die promptly and prevent further spread of infection, resistance.

- **hypertonia**
See **hypertonicity.**

- **hypertonic**
a solution whose osmotic potential is less than that of living cells, causing water loss, shrinkage or plasmolysis of cells.

- **hypertonicity**
See **hypertonic.**

- **hypertrophy**
the enlargement of tissues by an increase of the size of the cells hyperplasia.

- **hypervariable locus**
locus with many alleles, especially those whose variation is due to variable numbers of tandem repeats.

- **hypervariable region**
the part of a variable region that actually determines the specificity of an immunoglobulin.

- **hypha**
a tubular, threadlike filament of fungal mycelium.

- **hypocotyl**
part of the embryonic shoot or seedling located below the cotyledon and above the radicle.

- **hypogeal**
living or growing underground.

- **hypogean**
hypogeal.

- **hypomorph**
a mutant with less than the normal amount of some gene product.

- **hypoplasia**
abnormal deficiency of cells or structural elements.

- **hypoploid**
aneuploid with a small number of chromosomes missing.

- **hypoploidy**
  having missing chromosome complements compared to the standard chromosome set.

- **hypostatic epistasis**
  See **epistasis.**

- **hypostatic gene**
  a gene whose expression is masked by an epistatic gene.

- **hypotonic**
  of or designating a solution of lower osmotic pressure than another, as opposed to hypertonic.

- **i1, i2, i3, etc.**
  the first, second, third, etc., generations obtained by inbreeding.

- **IAA**
  Indole-Acetic Acid.

- **ichthyosis**
  any of several hereditary or congenital skin conditions. Skin of affected individuals has a dry, scaly appearance.

- **identical by descent**
  two genes that are identical in nucleotide sequence because they are both derived from a common ancestor.

- **identical in structure**
  two genes that are identical in nucleotide sequence, regardless of whether or not they are both derived from a common ancestor.

- **identity by descent**
  the state of two alleles when they are identical copies of the same ancestral allele (autozygous.)

- **Identity Preservation (IP)**
  a system of crop or raw material management that preserves the identity of the source or nature of the materials.

- **ideotype**
  crop plant with model characteristics known to influence photosynthesis, growth and grain production.

- **ideotype breeding**
  a method of breeding to enhance genetic yield potential based on modifying individual traits where the breeding goal for each trait is specified.

- **idioblast**
  a plant cell committed to develop into a cell type that differs from the surrounding tissue.

- **idiochromosome**
  a chromosome that contributes to the determination of sex.

- **idiogamy**
  combination of male and female gametes from the same individual.

- **idiogram**
  a diagrammatic representation of the karyotype of a plant.

- **idioplasm**
  all hereditary determinants of a plant including genotype and plasmotype germplasm.

- **idiotype**
  the sum of the hereditary determinants of a cell or plant consisting of the genotype and plasmotype.

- **idiotypic variation**
  variation unique to an individual immunoglobulin molecule produced by the variable region of immunoglobulin genes.

- **idling reaction**
  the production of guanosine tetraphosphate (3' ppGpp 5', magic spot) by the stringent factor when a ribosome encounters an uncharged tRNA in the A site.

- **ig**
  see **immunoglobulin**.

- **ile**
  isoleucine (an amino acid).

- **I-LINE**
  Inbred Line.

- **illegitimate crossing-over**
  unequal crossing-over.

- **illegitimate recombination**
  recombination between DNA fragments that do not share extensive DNA sequence homology. Transposons and insertion sequences have special functions that catalyse illegitimate recombination.

- **image processing**
  various mathematical procedures to improve the signal-to-noise and contrast and to obtain quantitative intensity data from images.

- **imbibition**
  the adsorption of liquid, usually water, into ultramicroscopic spaces or pores found in material such as cellulose, pectin and cytoplasmic proteins in seeds.

- **imidazole**
  a compound whose molecule forms a pentagonal ring of C and H atoms with an N and NH group attached.

- **imino acid**
  an acid derived from an imine in which the nitrogen of the imino group and the carboxyl group are attached to the same carbon atom.

- **immature**
  not mature or ripe.

- **immersed**
  embedded in a substrate.

▪ **immersion medium**
material placed between the uppermost surface of a microscopic sample (slide) and the objective (e.g., immersion oil).

▪ **immigration**
in genetics, the movement or flow of genes into a population, caused by immigrating individuals, which interbreed with the residents.

▪ **immobilisation**
the conversion of a chemical compound from an inorganic to an organic form as a result of biological activity.

▪ **impeder**
an individual of any value actually impeding the development of another individual of higher grade.

▪ **imperfect flower**
unisexual flowers, flowers lacking either male or female parts.

▪ **imperfect state (of fungi)**
the asexual state of a fungus (i.e., the state in which no sexual reproduction occurs).

▪ **implant**
material artificially placed in an organism.

▪ **imprinting**
see **molecular imprinting**.

▪ **imprinting**
a chemical modification of a gene allele which can be used to identify maternal or paternal origin of chromosome.

▪ **improvement planting**
any planting done to improve the value of a stand and/or experiment and not to establish a regular plantation.

▪ **in situ**
in place, where naturally occurring.

▪ **in situ conservation**
literally 'on-site' conservation. The conservation of plants or animals in areas where they developed their distinctive properties in the wild or in farmers' fields. Compare to *ex situ* conservation.

▪ **in situ hybridisation**
use of a DNA or RNA probe to detect the presence of the *complementary DNA* sequence in cloned bacterial or cultured *eukaryotic* cells.

▪ **in vitro**
outside a living organism.

▪ **in vitro collection**
1. a collection of germplasm maintained as plant tissue grown in active culture on solid or in liquid medium, it can be maintained as plant tissue rang-

ing from protoplast and cell suspensions to callus cultures, meristems, shoot-tips and embryos.
2. the cell, organ or tissue culture performed under artificial conditions in tubes, glasses, dishes, etc.

- **in vitro fertilisation**
pollination performed aseptically in vitro by direct application of the pollen to the ovule, it is used to overcome prezygotic incompatibility.

- **in vitro marker**
a mutation that allows identification in vitro of a cell line possessing the marker.

- **in vitro mutagenesis**
1. the production of either random or specific mutations in a piece of cloned DNA. Typically, the DNA will then be reintroduced into a cell or an organism to assess the results of the mutagenesis.
2. methods for altering DNA outside the host cells, mutagenesis can be random or specific for the site and base change depending on the technique used.

- **in vitro pollination**
in vitro fertilisation.

- **in vitro propagation**
propagation of plants under a controlled and artificial environment, usually applying plastic or glass vessels, aseptic techniques and defined growth media.

- **in vitro screening**
search and selection for particular characters of cells, organs or tissues performed under artificial conditions in tubes, glasses, dishes, etc., usually in combination with special nutritional media, which allow a differentiated growth of the cells, etc.

- **in vivo**
literally, 'in life,' applied to studies and propagation of whole, living organisms, on intact organ systems therein or on populations of micro-organisms.

- **inarable**
a field or land not arable and/or not capable of being ploughed or tilled.

- **inarching**
a method of grafting, usually a new plant growth onto a stronger root system. It is carried out by establishing young plants (sometimes, one that is in a pot) near an existing tree, at the point where they meet. At the matching areas the bark is removed, the two cut surfaces are then fitted together and bound with soft tying material until they grow together, later they

can be gradually separated with the new branches attached to the older rootstock.

- **inbred**
  a plant resulting from successive self-fertilisation of parents throughout several generations.

- **inbred line**
  a line produced by continued inbreeding, usually a nearly homozygous line originating by continued self-fertilisation, accompanied by selection.

- **inbred pure lines**
  involves inbreeding of annual seed-propagated material, homogeneous, homozygous isolated by selection of desired recombinants or segregates in F2 to F7 generations of crosses between parental pure lines (generally monogenotypic, can be blended to form multilines, e.g., tomato, lettuce, soybean, pea, cowpea, snapbean, field bean, Arabian coffee, *Capsicum* pepper, eggplant, okra, lentil and papaya.)

- **inbred-variety cross**
  the F1 cross of an inbred line with a variety toppers.

- **inbreeding**
  the crossing of closely related plants. One important purpose of induced inbreeding is the development of genotypes that can be maintained through multiple generations of seed production. Self-pollinated cultivars are reproduced for many generations by inbreeding. Inbreeding is also used to reduce the frequency of deleterious recessive alleles in genotypes that serve as parents of a synthetic or a vegetatively propagated cultivar. Inbreeding increases the genetic and phenotypic variability among individuals in a population. Four mating systems are used to increase the homozygosity in a breeding population.

- **inbreeding coefficient**
  the probability that the two genes at any locus in a diploid individual are identical by descent (i.e., they originated from the replication of one gene in a previous generation.)

- **inbreeding depression**
  the reduction in vigour often observed in progeny from matings between close relatives. It is due to the expression of recessive deleterious alleles it is usually severe in open-pollinated outcrossing species, an effect opposite to heterosis heterosis.

- **inbreeding load**
  the extent to which a population is impaired by inbreeding.

■ **incertae sedis**
of uncertain taxonomic position.

■ **incipient species**
populations that are too distinct to be considered as subspecies of the same species, but not sufficiently differentiated to be regarded as different species, sometimes called 'semispecies'.

■ **inclined draper**
a device for separating seeds using an inclined endless belt onto which seeds are metered. Seeds are separated on the basis of their different tendencies to roll down the plane or to catch and be carried up and into a separate discharge spout.

■ **inclusive fitness**
the expansion of the concept of the fitness of a genotype to include benefits accrued to relatives of an individual since relatives share parts of their genomes. Hence an apparently altruistic act toward a relative may in fact enhance the fitness of the individual performing the act.

■ **incompatibility**
a genetically determined inability to obtain fertilisation and seed formation after self-pollination or cross-pollination. There are several types of progamous or postgamous incompatibility, in contrast to heteromorphic incompatibility (e.g., heterostyly in *Primula* spp.), homomorphic incompatibility is not associated with morphological differences.

■ **indehiscent fruit**
indehiscent.

■ **indent cylinder separator**
a seed separator utilising a rotating indented cylinder through which seeds are passed for cleaning, it lifts shorter seeds from longer seeds, thus separating them.

■ **indicator plant**
plants that are indicative of specific site or soil conditions.

■ **indoor culture**
growing plants indoors using natural and/or artificial light and additional heating, it is used for subtropical or tropical plants or for plant propagation.

■ **indoor plant**
See **indoor culture**.

■ **induced mutation**
a change in a gene caused by a treatment.

■ **induction of flowering**
the initiation of the production of flowers, possibly stimulated by florigen.

■ **inert**
a chromosomal segment that is supposed to be genetically inactive or without coded genetic information.

■ **infect**
of a pathogen, to enter and establish pathogenic relationship with an organism, to enter and persist in a carrier, to make an attack on a plant.

■ **infection**
the invasion of the tissue of a plant by a pathogenic microorganism.

■ **inferior pelea**
lemma.

■ **infertility**
the situation in which a plant is unable to produce viable offspring.

■ **infest**
attacked by animals (e.g., insects) or sometimes used of fungi in soil in the sense of contaminated.

■ **inflected**
when the keel of the, for example, wheat glume is bent inward in the upper third.

■ **inflorescence**
a flower structure that consists of more than a single flower, the flower head terminates the culm in grasses. It may be determinate (solitary flower, simple cyme, compound cyme, scorpioid cyme, glomerule) or indeterminate (raceme, panicle, spike, catkin, spadix, umbel, head). Determinate flowers are those in which the axis terminates as a flower, indeterminate flowers terminate in a bud, which continues to grow and produce flowers throughout the growing season, the latter results in flowers of different maturity within the same inflorescence.

■ **infructescence**
a fruiting structure that consists of more than a single fruit.

■ **infundibular**
funnel-shaped.

■ **ingraft**
to insert, as a scion of one tree or plant into another, for propagation.

■ **initiation codon**
the mRNA sequence AUG, which specifies methionine, the first amino acid used in the translation process. (Occasionally GUG, valine, is recognised as an initiation codon).

■ **initiation complex**
the complex formed for initiation of translation. It consists of the 30S ribosomal subunit, mRNA, N-formyl-methionine tRNA and three initiation factors.

- **initiation factors (if1, if2, if3)**
  proteins (prokaryotic with eukaryotic analogues) required for the proper initiation of translation.

- **initiator protein**
  protein that recognises the origin of replication on a replicon and takes part in primosome construction.

- **inoculant**
  a preparation containing specific nitrogen-fixing bacteria that is added to legume seed prior to planting to assure that the resulting crop will have nitrogen fixation ability.

- **inoculation**
  the act or process of inoculating , in agriculture, addition of effective *Rhizobia* (bacteria) to legume seed prior to planting for the purpose of promoting nitrogen fixation inoculate.

- **inositol**
  a carbocyclic or sugar alcohol that is widely distributed in plants.

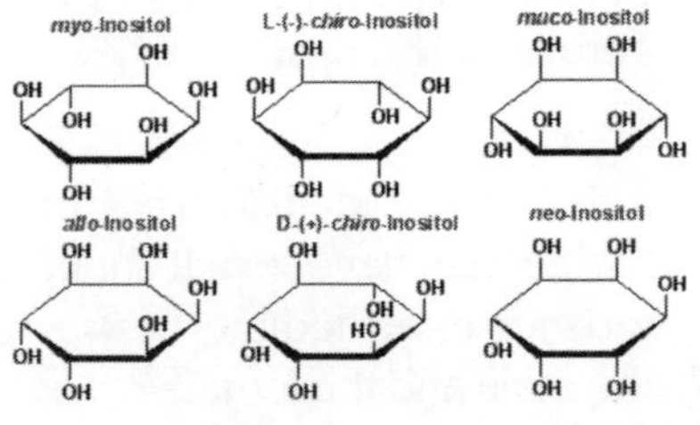

- **inositol triphosphate (insp3)**
  a chemical compound that contributes to the spatial orientation of a plant, for example, when wheat or maize plants are pressed down to the ground, a change of orientation of starch granules occurs in the cells, after a short time (30-120 minutes) InsP3 is accumulated on the lower side of the leaves, so-called motor cells are activated, they grow longitudinal and stepwise upright the plant.

- **insect herbivory**
  plant damage by certain insects (e.g., fall armyworm, *Spodoptera frugiperda*), it is responsible for about 15 % of the world's crop losses each year.

- **insecticide**
  a substance or preparation used for killing insects.

- **intake auger (at a harvester)**
  the auger tines guide the crop to the chain conveyor, which delivers it to the threshing section, any foreign bodies that may have been ingested fall into the stone trap, which is located between the conveyor and the concave.

- **integrated plant protection**
  disease and pest control by combining all available techniques, such as agronomic con-

trol, biological control, chemical control and sanitary procedures.

- **interbreeding**
intercrossing of individuals within a population.

- **intercropping**
two or more crops produced on the same field at the same time.

- **intergeneric cross**
spontaneous or experimental crosses of individuals of different genera, for example, wheat *(Triticum aestivum)* and rye *(Secale cereale)*. This cross even resulted in a human-made new crop plant 'triticale'.

- **interkinesis**
the abbreviated interphase that occurs between meiosis I and II. No DNA replication occurs here.

- **interlocking**
during meiotic pairing, the intertwisting of nonhomologous chromosomes and/or chromosome configurations.

- **intermediary**
a plant trait controlled by a heterozygous pair of alleles, which result in an intermediate phenotype as compared to the corresponding homozygous genotypes.

- **internal hair (of the glume)**
the hair situated across the upper part of the internal surface of the broad wing in the glumes of, for example, wheat.

- **internal inprint**
the mark on the inner surface of, for example, wheat glume caused by the pressure of the enclosed lemma and grain.

- **internode**
the part of a stem between two consecutive nodes.

- **interphase**
the period in the cell cycle when DNA is replicated in the nucleus, followed by mitosis.

- **interplot competition**
it can be avoided and/or decreased by use of plots with multiple rows in which only plants in the centre rows are evaluated. In plots with three or more rows, the outermost rows are designated as the border or guard rows, they may prevent plants in adjacent plots from influencing the performance of plants in the centre of the plot.

- **intersex**
a class of individuals of a bisexual species that have sexual characteristics intermediate between the male and the female.

- **interspecific hybrid**
  a hybrid between two or more species.

- **intragenic**
  effects and phenomenons within a gene or its physical unit.

- **intragenic suppressor**
  a mutation that suppresses the phenotype of another mutation in the same gene as that in which the suppressor mutation resides.

- **intraspecific**
  effects and phenomenons within a species.

- **intravarietal**
  effects and phenomenons within a variety (cultivar).

- **introns**
  a segment of DNA (between exons) that is transcribed into nuclear RNA, but are removed in the subsequent processing into mRNA.

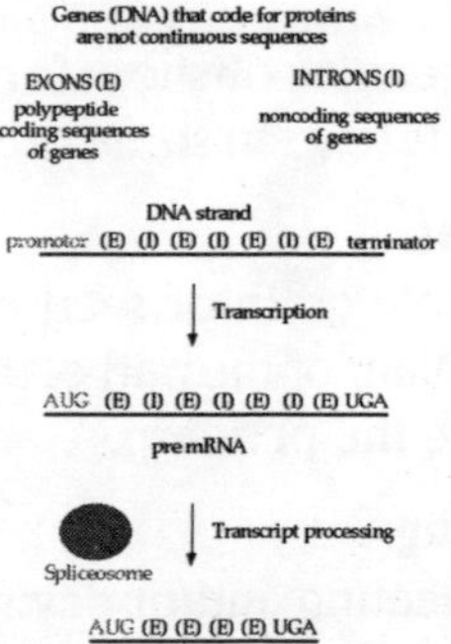

- **invasion**
  the spreading of a pathogen through tissues of a diseased plant.

- **inversion**
  a change in the arrangement of genetic material involving the excision of a chromosomal segment that is then turned through 180° and reinserted at the same position in the chromosome.

- **inviability**
  the inability to survive.

- **involucre**
  a whorl of bracts below an inflorescence.

- **involute**
  having edges that roll under or inwards.

- **iojap**
  an idiomatic description of a mutant locus in maize that produces variegation.

- **irrigation**
  to supply land with water by artificial means, as by diverting streams, flooding or spraying, main types of irrigation are (1) sprinkler irrigation, (2) surface irrigation, (3) subsurface irrigation.

- **isoallele**
  an allele whose effect can only be distinguished from that of the normal allele by special tests.

- **isocompetition**
  cultivation at high plant density implies the presence of strong interplant competition, when there is no genetic variation, the competition is called isocompetition.

- **isoenzyme**
  a species of enzyme that exists in two or more structural forms, which are easily identified by electrophoretic methods.

- **isogamete**
  male and female gametes that are similar to each other.

- **isolation requirement**
  the spatial separation required between a seed field and other sources of mechanical and genetic contamination, especially between cross-pollinated varieties.

- **joining segment**
  a small DNA segment that links genes to yield a functional gene encoding an immunogobulin.

- **joint**
  node.

- **jointing stage**
  in cereals, the growth stage at which the first stem node is visible above ground.

- **jumbo pollen**
  it refers to a pollen trait in diploid alfalfa, homozygous recessive (*jpjp*) plants are charactetised by the complete failure of post-meiotic cytokinesis during microsporogenesis resulting in 100% 4n-pollen formation.

- **kappa particle**
  bacteria-like particle that gives a *Paramecium* the killer phenotype.

- **keel**
  the main nerve of, for example, the wheat glume, shaped somewhat like a keel of a boat, in legumes also a boatlike formation of the flower.

- **keel flower**
  boatlike shape of a flower (e.g., in legumes such as pea.)

- **keiki**
  a vegetative offshoot formed at a node (e.g., in some orchids.)

- **kernel**
  a whole grain or seed of a cereal plant or the part of the seed inside the pericarp.

- **kilning**
  the heating and/or drying process used in the production of

malt to stop germination and kill the grain.

- **kjeldahl method**
  a technique often used for the quantitative estimation of the nitrogen content of plant material (e.g., of cereal grains).

- **knot**
  a lump or swelling in or on a part of a plant (e.g., the node of grass).

- **labellum**
  in Orchidaceae, the lowest of the three flower petals, which differs from the other two, in lipped flowers, the platform formed by the lowest petal or fused petals.

- **lamina**
  a flat, sheetlike structure (e.g., the blade of a leaf.)

- **late crop**
  a crop or plant showing late maturation within a given season.

- **late wood**
  a result of secondary plant growth, very young branches have hardly any fascicular cambium, interfascicular cambium develops very early during the year, even before the beginning of secondary growth, the activity of the cambium increases branch diameter and the vascular bundles become elongated in cross section, far more xylem than phloem elements are produced. Annual rings become clearly visible because at the beginning of each vegetation period (in spring) vessels (conducting function) and fibers (supporting function) with a wide lumen are assembled first, the so-called early wood, in the following season. Elements with steadily narrowing volumes are produced, in autumn. Only a few vascular elements with narrow lumina (late wood) form.

- **lateral nerve(s)**
  for example, in wheat, the nerves, which run along the length of the broad and narrow wings of the glume, in barley, the two pairs of nerves (inner and outer) lying toward the margins of the lemma and on either side of the median nerve.

- **lateral root**
  roots arising from the main root axis.

- **lateral shoot**
  shoots originating from vegetative buds in the axils of leaves or from the nodes of stems, rhizomes or stolons.

- **late-sown**
  sowing date later than the optimal time for a given crop or variety.

- **latex**
  a white, commonly sticky substance produced in specialised tissues within a plant.

- **latifoliate**
  broad-leafed.

- **layering**
  covering stems, runners or stolons with soil causing adventitious roots to form at the nodes, enabling propagation by rooted cuttings, this procedure is used commercially to propagate many plants, in vitro layering, the horizontal placement of cultured shoots or nodal segments on agar growth medium in order to produce axillary bud formation.

- **leaching**
  the washing out of material from the soil, both in solution or suspension.

- **leaf**
  a thin, usually green, expanded organ born at a node on the stem of a plant, typically comprising a petiole (stalk) and blade (lamina) and subtending a bud in the axil of the petiole, it is the main site of photosynthesis.

- **Leaf Area Index (LAI)**
  the total leaf surface area exposed to incoming light energy, expressed in relation to the ground surface area beneath the plant (e.g., LAI = 3, the leaf area exposed to light is three times of the ground surface area).

- **leaf axil**
  the angle between a petiole and the stem.

- **leaf bud**
  a bud producing a stem and leaf, unlike a flower bud, which contains a blossom.

- **leaf curling**
  a trait that is expressed in several plants caused either by abiotic stress, diseases or mutant genes, e.g., the soft red winter wheat varieties 'Kaskaskia' or 'Penjamo 62' (registered in USA) exhibit leaf rolling, this trait is pronounced under some heat or drought stress, the expression is most visible just prior to heading, the flag leaves curl up lengthwise or some triticales that tightly roll the leaves to give the appearance of an onion leaf.

- **leaf cutting**
  a cutting made from a single leaf, a method for propagation (e.g., of succulents), a leaf can be knocked off or cut off the plant, either it spontaneously roots on the ground or it is

placed in certain media for rooting.

- **leaf fall**
  leaf abscission.

- **leaf posture**
  the characteristic position of the foliage leaves on the stem axis, which imprints the plant habit of the species, it may contribute to optimal utilisation of light and thus photosynthesis, there were several approaches to breed for specific leaf posture in order to improve photosynthetic capacity of, for example, cereals.

- **leaf primordium**
  a lateral outgrowth from the apical meristem that develops into a leaf.

- **leaf sheath**
  a tubular envelope, as the lower part of the leaf in grasses.

- **leaf vein**
  vascular bundles in the leaves, in the petiole and the midvein of the leaf the veins are very large. Farther out into the mesophyll the veins may consist of only one xylem or phloem element, in these regions, these very small veins are called veinlets.

- **leafstalk**
  the footstalk or supporting stalk of a leaf petiole.

- **lectin**
  a generic term for proteins extracted from plants (e.g., legumes) that exhibit antibody activity in animals.

- **leghaemoglobin**
  an iron-containing, red pigment produced in root nodules during the symbiotic association between rhizobia and leguminous plants legume(s).

- **legume(s)**
  plants showing a simple or single pistil and characterised by a dry fruit pod that splits open by two longitudinal sutures and has a row of seeds on the inner side of the ventral suture (e.g., bean, pea, soybean, locust), there are many valuable food, forage and cover species, such as peas, beans, soybeans, peanuts, clovers, alfalfas, sweet clovers, lespedezas, vetches and kudzu, sometimes referred to as nitrogen-fixing plants, legumes are an important rotation crop because of their nitrogen-fixing property.

- **lemma**
  flowering glume, the lower or outer of the two bracts of the floret.

- **lesion**
  a visible area of diseased tissue on an infected plant.

- **leucoplast**
  a colorless plastid that is involved in the metabolism and storage of starches and oils.

- **lid**
  the cap of a boxlike seed capsule.

- **ligand**
  an atom, ion or molecule that acts as the electron donor partner in one or more co-ordination bonds or a molecule (e.g., antibody), which can bind to specific sites on cell membranes.

- **light soil**
  a soil that has a coarse texture and is easily cultivated.

- **lignification**
  converting into wood, cause to become woody.

- **ligula**
  a scalelike membrane that covers the surface of a leaf, in some Compositae, a strap-shaped corolla, sometimes, a fringe of epidermal tissue found at the boundary between the sheath and the blade of a maize leaf 30.

- **ligule**
  ligula.

- **lime**
  compounds of calcium used to correct the acidity in soils.

- **line breeding**
  a system of breeding in which a number of genotypes, which have been progeny tested in respect of some characters, are composited to form a variety, examples of line varieties are from normally self-fertilised crops, e.g., 'Gaines' wheat, line.

- **lining out**
  transplanting seedlings or rooted cuttings in rows in a nursery bed.

- **linola**
  a new form of linseed known by the generic crop name 'solin,' which produces a high-quality edible polyunsaturated oil similar in composition to sunflower oil.

- **lint (linters)**
  the long fibres of cotton seed. The short fibres generally remain attached to the seed in ginning, sometimes called 'fuzz', they are used mainly for batting, mattress stuffing and as a source of cellulose. Presence or absence of lint and fuzz are controlled by the interaction of four gene loci on non-homologous chromosomes, these loci were designated as *N1, N2, Li3* and *Li4*, where *N1 N1* confers the presence of fuzzy, *N2 N2* confers inhibition of fuzzy initiation and

development and duplicate gene pairs, *Li3Li3* and *Li4Li4*, determine the presence of lint, homozygosity for *li3li3* and *li4li4* might also inhibit fuzz from development, in other words, they were recessive epistatic to fuzz genes (DU, X. M. et al., 2001).

- **l-notch planting**
  a form of slit planting involving two slits at right angles with the seedling placed at the apex of the 'L'.

- **loam soil**
  a soil containing sand, silt and clay.

- **lodging resistance**
  plants that can resist lodging by optimal root system, stiffer straw or other characteristics (e.g., in cereal breeding, the introduction of 'semi-dwarf genes' contributed to shorter plants and thus higher lodging resistance even when nitrogen fertilisation is increased) lodging near-isogenic lines semidwarf *Rht* gene.

- **lodiculae**
  two small, translucent, scale-like structures situated at the base of the floret.

- **loess (soil)**
  unconsolidated, wind-deposited sediment composed largely of silt-sized quartz particles (0.015-0.05 mm diameter) and showing little or no stratification.

- **loment(um)**
  a dry schizocarpic fruit in the form of a legume or siliqua with constrictions formed between the seeds as it matures, so that the final fruit is composed of one-seeded, indehiscent loment segments.

- **long-day plant**
  a plant in which flowering is favoured by long days (>14 h daylight) and corresponding short dark periods, there are two types: species in which there is an absolute requirement for these conditions and others in which flowering is merely hastened by them.

- **longevity**
  the persistence of an individual for longer than most members of its species or of a genus and/or species over a prolonged period of geological time.

- **long-term storage**
  storage of seeds in a gene bank longer than ten years.

- **lopping**
  a procedure by which all the branches of a tree are cut off, except the leading shoot, as opposed to pruning, in which only some of the branches are cut.

- **low-input variety**
  a crop variety with low claims at macro- and micronutrient fertilisers, pest control and agronomic measures.

- **lunate**
  shaped like a half-moon.

- **luxuriance**
  hybrids that are larger, faster growing or otherwise exceed the parental forms in some traits. It is usually brought by complementary gene action present in the parents and combined in the hybrid heterosis.

- **lysozyme**
  an enzyme that is destructive of bacteria and functions as an antiseptic, found in certain plants.

- **M2 population**
  the progeny derived from selfing M1 plants, which themselves are progeny that arise by selfing plants grown from mutagenised seed, recessive mutations, resulting from the seed mutagenesis, are detected in M2 plants, which are homozygous for the mutation.

- **maceration**
  softening of plant tissue by using of enzymes, hydrolic acid or other means, usually, the middle lamella of the cell walls is degraded without modification of the cell content.

- **macerozyme**
  an enzyme or a mixture of enzymes able to soften plant tissue maceration.

- **macrocarpous**
  carrying or forming big fruits.

- **macroclimate**
  the general climate of a large area, as that of a continent or country, as opposed to microclimate.

- **macroelement**
  chemical elements, such as nitrogen or phosphorus, that are needed in large amounts as nutrients for plant growth.

- **macroevolution**
  evolution above the species level (i.e., the development of new species, genera, families, orders, etc.)

- **macromolecule**
  a large polymer such as DNA, RNA, protein, lipid or polysaccharide.

- **macromutation**
  a mutation that results in a profound change in an organism, as a change in a regulatory gene that controls the expression of many structural genes, as opposed to micromutation.

macronutrient

an inorganic element or compound that is needed in relatively large amounts by plants.

macrorestriction map

map depicting the order of and distance between sites at which *restriction enzymes* cleave *chromosomes*.

macrostylous

showing long stamen.

maculate

spotted or blotched.

magnesium

an element that is found in high concentrations in plants, it plays an important role in the chemical structure of chlorophyll and of membranes and is involved in many enzyme reactions, especially those catalysing the transfer of phosphate compounds, deficiency can produce various symptoms, including chlorosis and the development of other pigments in leaves.

maize gluten

a byproduct of wet milling, it is used as a medium-protein (20-24 percent) and medium-fibre (10 percent) foodstuff.

major gene

a gene with pronounced phenotypic effects, in contrast to modifier gene, which modifies the phenotypic expression of another gene oligogene.

male sterility

producing no functional pollen.

malformation

faulty or anomalous formation or structure.

malt

germinated grain used in brewing and distilling.

maltose

a disaccharide that consists of two alpha-glucose units linked by an alpha-1,4-glycosidic bond.

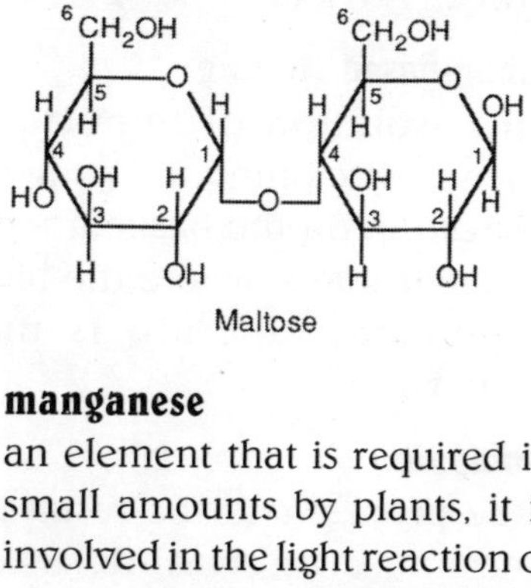

Maltose

manganese

an element that is required in small amounts by plants, it is involved in the light reaction of photosynthesis and also binds proteins, deficiency causes interveinal chlorosis and malformation.

mannitol

a polyhydroxy alcohol that can be synthesised chemically by the reduction of mannose and is present in many plants.

- **mannose**
a hexose, $C_6H_{12}O_6$, obtained from the hydrolysis of the ivory nut and yielding mannitol upon reduction.

- **manure**
animal excreta with or without a mixture of bedding or litter.

- **map unit (m.u.)**
the distance between two linked gene pairs where 1 percent of the products of meiosis are recombinant. A unit of distance in a linkage map. The distance equal to 1% recombination between two loci.

- **map-based cloning**
the isolation of important genes by cloning the gene in question on the basis of molecular maps, where the biochemical function is unknown.

- **mapping**
the process and the result of determination of map distances within or between linkage groups, sometimes it refers to the localisation of genes or chromosome segments physical map.

- **marginal farmland**
land repeatedly farmed without benefit of humus or chemical replacements.

- **marker**
an identifiable physical location on a chromosome (e.g., restriction enzyme cutting site,) whose inheritance can be monitored. Markers can be expressed regions of DNA (genes) or some segment of DNA with no known coding function but whose pattern of inheritance can be determined.

- **masked symptoms**
plant symptoms (e.g., caused by a virus) that are absent under some environmental conditions but appear when the host is exposed to certain conditions of light and temperature.

- **mass emasculation**
in hybrid breeding (e.g., in maize), the emasculation of the male flowers by mechanical means emasculate detasseling.

- **mass selection (positive or negative)**
a form of breeding in which individual plants are selected on their individual advantages and the next generation propagated from the aggregate of their seeds. The easiest method is to select and multiply together those individuals from a mixture of phenotypes, which correspond to the breeding aim (positive mass selection), it is still applied in cross-pollinating

of vegetable species, such as carrots, radishes or beetroots, in order to improve the uniformity, when all undesired off-types are rouged in grown crop population and the remaining individuals are propagated further. The method is termed negative mass selection. Negative mass selections is no longer an adequate breeding method for highly advanced varieties, it is usually applied in multiplication of established varieties (i.e., for seed production in order to remove diseased plants, casual hybrids or other defects).

- **mating design**
the pattern of pollination set up between individuals for an artificial crossing programme, e.g., single pair mating, double pair mating there each parent participates in one or two crosses.

- **mating group**
a group of individuals that gives the chance for mating among one another on the basis of genetic prerequisites.

- **mating system**
the pattern of mating in sexually reproducing organisms, two types of mating systems are: (1) random mating and (2) assortative mating (genetic assortative mating, genetic disassortative mating, phenotypic assortative mating, phenotypic disassortative mating).

- **mating type**
the genetic properties of an individual for a particular type of mating.

- **matroclinal**
with hereditary characteristics more maternal than paternal (e.g., in certain banana hybrids.)

- **maturation**
the completion of development and the process of ripening.

- **maturation division**
meiosis.

- **mature**
fully differentiated and functionally competent cells, tissues or organisms.

- **mature resistance**
adult resistance.

- **maysin**
a flavone glycoside present in silks of maize, the adoption of high-pressure liquid chromatography procedures allowed the identification of analogues, such as (apimaysin (AP), 3-methoxymaysin (ME), isoorientin and other luteolin derivatives) as well as other compounds, such as chlorogenic acid (CHA), which demonstrated antibiotic activity

to the maize earworm, the generally accepted level of maysin in silks required to reduce maize earworm larval weights by 50% is approximately 0.2% of silk fresh weight.

- **MDNA**
messenger DNA.

- **mechanical inoculation**
a method of transmitting the pathogen from plant to plant, for example, sap from diseased plants or a defined inoculum are rubbed on test-plant leaves that usually have been dusted with carborundum or other abrasive materials. It is applied in experimental testing of plant resistance.

- **media composition**
different supplements to the nutritive substance provided for the growth of a given plant in the laboratory.

- **median centromere**
a centromere that is located midway of chromosomes resulting in two equal-long arms.

- **medical plants**
plants that are or have been used medicinally (e.g., chamomile).

- **medium**
any material on (or in) which experimental cultures are grown.

- **medium-term seed storage**
with a storage time of about ten years long-term storage.

- **mega-environment (ME)**
a broad, not necessarily contiguous area, occurring in more than one country and frequently transcontinental, defined by similar biotic and abiotic stresses, cropping system requirements, consumer preferences and by a volume of production.

- **megagametogenesis**
the development of the female gametophyte from a functional megaspore.

- **megagametophyte**
embryo sac.

- **megaspore**
one of the four cells formed in the ovule of higher plants as a result of meiosis or sexual cell reduction division, one of these later undergoes mitosis to give rise to the female gamete.

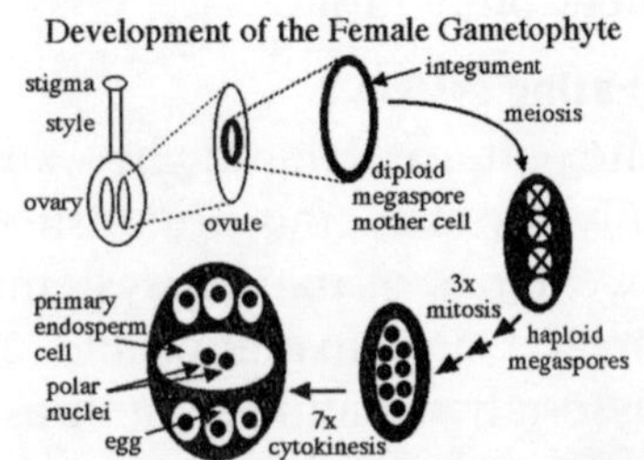

- **megaspore mother cell**
a diploid cell in the ovary that gives rise, through meiosis, to four haploid megaspores.

- **megasporocyte**
embryo sac.

- **megasporogenesis**
the development of the megaspore from the archesporial cell.

- **mega-yeast artificial chromosomes (mega yac)**
a large (>500 bp) piece of DNA that has been cloned inside a living yeast cell. While most bacterial vectors cannot carry DNA inserts that are larger than 50 bp and standard YACs typically cannot carry DNA pieces that are larger than 500 bp, mega YACs can carry DNA pieces (chromosomes) as large as one million bp.

- **meiocyte**
cell in which meiosis occurs.

- **meiosis**
1. two successive nuclear divisions (with corresponding cell divisions) that produce haploid gametes (in animals) or haploid sexual spores (in plants and fungi) having one-half of the genetic material of the original cell. The nuclear and cell division process in diploid eukaryotes that results in four haploid gametes or spores having one member of each original pair of homologous chromosomes only per nucleus.
2. a type of nuclear division that occurs at some stage in the life cycle of sexually reproducing organisms, by a specific mechanism the number of chromosomes is halved to prevent doubling in each generation, genetic material can be exchanged between homologous chromosomes, there are different stages: leptotene, zygotene, pachytene, diplotene, diakinesis, metaphase I, anaphase I, telophase I, interphase, metaphase II, anaphase II, telophase II, microspore formation.
3. the doubling of gametic chromosome number.

- **meiospore**
cell that is one of the products of meiosis in plants.

- **meiotic crossing-over**
chiasma.

- **meiotic cycle**
See **meiosis.**

- **meiotic duration**
the time needed for completing the cell cycle from prophase to telophase under certain conditions.

- **melting of DNA**
denaturation of duplex DNA by heat or increased pH leading to strand separation.

- **membrane**
a sheetlike structure, 7-10 nm wide, that forms the boundary between a cell and its environment and also between various compartments within the cell, it is composed of lipids, proteins and some carbohydrates. It functions as a selective barrier and also as a structural base for enzymes.

- **Mendel's laws of inheritance, Mendelian laws of inheritance**
the inheritance of chromosomal genes on the basis of chromosome theory of heredity, three laws are considered: (1) law of dominance or of uniformity of hybrids (2) law of segregation (3) law of independent assortment.

- **Mendelian character**
a character that follows the laws of inheritance formulated by G. Mendel.

- **Mendelian inheritance**
Mendel's laws of inheritance.

- **Mendelian population**
an interbreeding group of organisms that share a common gene pool.

- **Mendelian ratio**
the segregation rations according to Mendel's laws of inheritance.

- **Mendelism**
Mendel's laws of inheritance.

- **Mendelise**
to segregate according to Mendel's laws of inheritance.

- **Mendel's first law**
the two members of a gene pair segregate from each other during meiosis, each gamete has an equal probability of obtaining either member of the gene pair.

- **Mendel's second law**
the law of independent assortment, unlinked or distantly linked segregating gene pairs assort independently at meiosis.

- **mercaptan**
thiol.

- **mericlinal**
it refers to a chimera in which the inner tissue has a different genetic constitution than the surrounding outer tissue chimera.

- **meristem**
a group of plant cells that is capable of dividing indefinitely and whose main function is the production of new growth, meristematic cells are found at the growing tip of a root or a stem

(apical meristem), in cambium (lateral meristem) and also within the stem and leaf sheaths (intercalary meristem of grasses.)

- **meristem (tip) culture**
the culture of an explant consisting only of a meristematic part.

- **meristematic**
pertaining to the meristem.

- **meristematic tip**
the meristematic dome and one pair of leaf primordia, it is commonly used as explants, particularly to produce virus-free plant material meristem.

- **mesocarp**
middle layer of the fruit wall pericarp.

- **mesocotyl**
an elongated portion of the seedling axis between the point of attachment of the scutellum and the shoot apex of, for example, a grass seedling. It is recognised as a compound structure that is formed by the growing together of the cotyledon and the hypocotyl.

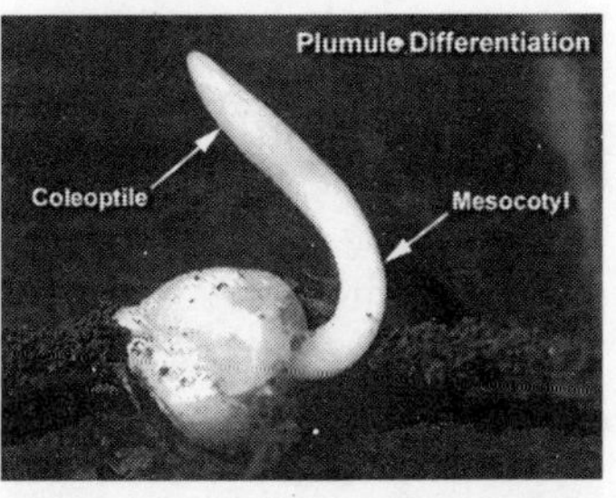

- **mesoderm**
the middle layer of embryonic cells between the ectoderm and the endoderm.

- **mesophyll**
internal parenchyma tissue of a plant leaf that lies between epidermal layers, it functions in photosynthesis and in storage of starch.

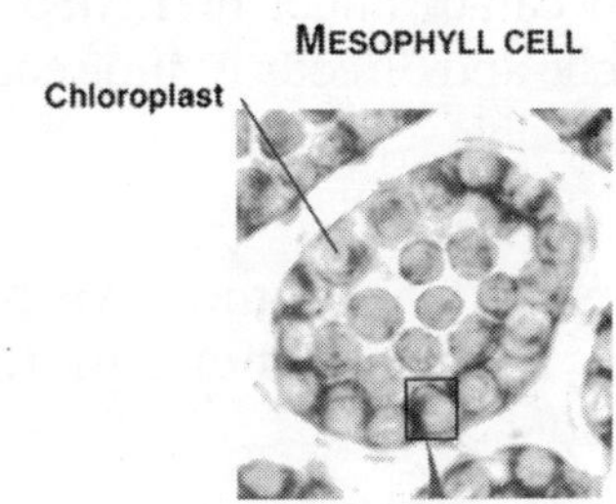

- **mesophyll explant**
an explant prepared from internal parenchyma tissue mesophyll.

- **mesophyte**
a plant with an intermediate water requirement.

- **messenger DNA (mDNA)**
a single-stranded DNA that acts as a messenger of the protein biosynthesis.

- **messenger RNA (mRNA)**
an RNA molecule transcribed from the DNA of a gene and from which a protein is translated by the action of ribosomes. The basic function of the nucleotide

sequence of mRNA is to determine the amino acid sequence in proteins.

- **met**
methionine (an amino acid).

- **metabolic pathway**
a sequential series of enzymatic reactions involving the synthesis, degradation or transformation of a metabolite. The pathway can be linear, branched or cyclic and directly or indirectly reversible.

- **metacentric**
applied to a chromosome that has its centromere in the middle.

- **metacentric chromosome**
a chromosome having its centromere in the middle.

- **metagon**
an RNA necessary for the maintenance of mu particles in Paramecium.

- **metamere**
segmental repeat unit in higher animals.

- **metaphase**
1. the stage of mitosis or meiosis in which spindle fibres are attached to kinetochores and the chromosomes are positioned in the equatorial plane of the cell. It is an intermediate stage of nuclear division when chromosomes align along the equatorial plane of the cell spindle.
2. a stage of mitosis or meiosis at which the chromosomes move about within the spindle until they eventually arrange themselves in its equatorial region. In metaphase I (MI) of meiosis, the chromosomes of a genome line up within the cell at a position referred to as the equatorial plate, spindle fibres form, which link each chromosome of a homologous pair to a different pole of the cell. The orientation of the chromosomes relative to the two poles seems to be random, the number of different combinations of chromosomes that can occur due to their orientation at MI is defined by the formula $2\ n-1$, where n is the number of chromosomes in the genome, for example, there are two combinations possible with two chromosome pairs, four combinations with three chromosome pairs and eight combinations with four chromosome pairs.

- **metaphase arrest**
the stopping of cell division at mitotic or meiotic metaphases, usually by application of specific agents.

■ **metaphase plate**
the plane of the equator of the spindle into which chromosomes are positioned during metaphase.

■ **metapopulation**
the some of multiple sublined breeding populations may be referred to as a metapopulation.

■ **metaxenia**
the influence of pollen on maternal tissue of the fruit.

■ **method of overstored seeds**
in pedigree breeding of allogamous crop plants (e.g., in rye), an effective method of regulating cross-fertilisation, usually, a greater number of individual plants is harvested from a genotypic mixture of a certain population, their progenies are sown as A families in smaller plots while half of the seed of all elite plants is retained in reserve, those A families meeting all the requirements are not directly multiplied but the remaining seed of the corresponding elite plants is sown in the following year. It enters a so-called A' family trial. In this way the economically valuable traits of the A family can be definitely evaluated after maturity, the best A families determined for further breeding have already been pollinated by a pollen mixture that also contains pollen of less valuable plants.

■ **methotrexate**
a toxic folic acid analogue, $C_{20}H_{22}N_8O_5$, that inhibits cellular reproduction.

■ **methylation**
modification of a molecule by the addition of a methyl group.

■ **methylene blue**
a vital staining agent for chromosomes.

■ **metric character**
a trait that varies more or less continuously among individuals, which are therefore placed into classes according to measured values of the trait, it is also called 'quantitative character'.

■ **microclimate**
the atmospheric characteristics prevailing within a small space macroclimate.

■ **microelement**
trace element macroelement micronutrient.

■ **microevolution**
evolutionary change within species that results from the differential survival of the constituent individuals in response to natural selection, the genetic variability on which the selection operates arises from muta-

tion and sexual recombination in each generation.

- **microgametogenesis**
  the development of the microgametophyte (pollen grain) from a microspore.

- **micromanipulation**
  manipulation or surgery done while viewing the object through a microscope and often carried out with the aid of an injection or dissection of substances or particles microinjection.

- **micromanipulator**
  the facility usually attached to a microscope in order to carried out micromanipulation micromanipulation.

- **micrometer**
  a unit of measurement frequently used in microscopy (1 $\mu m = 10^{-6}$ meter), in older usage also known as micron.

- **micronucleus**
  the smaller nucleus as distinguished from the larger nucleus, produced during telophase of mitosis or meiosis by lagging chromosomes or chromosome fragments derived from spontaneous or induced chromosome aberrations.

- **micronutrient**
  an inorganic element or compound that is needed in relatively small amounts by plants microelement trace element.

- **micro-organism**
  literally, a 'microscopic organism', the term is usually taken to include only those organisms studied in microbiology (bacteria, fungi, microscopic algae, protozoa, viruses.)

- **microphotography**
  photography requiring optical enlargement.

- **microplot**
  microplots are used to minimise the amount of seed or space required to evaluate a group of individuals, the number of plants in a microplot differs among crops, when short rows are used as microplots, the plant density is comparable to that of larger row plots, in unbordered microplots, the effect of interplot competition has to be considered when determining an appropriate distance among the plots.

- **microprojectile**
  high-velocity microprojectile transformation.

- **micropropagation**
  the in vitro culture or vegetative propagation of a plant in vitro propagation.

- **micropylar**

micropyle.

- **micropyle**

a canal in the coverings of the nucellus through which the pollen tube usually passes during fertilisation, later, when the seed matures and starts to germinate, the micropyle serves as a minute pore through which water enters.

- **microsatellite DNA**

pieces of small DNA sequences that are repeated (appear repeatedly in sequence within the DNA molecule) adjacent to a specific gene within the DNA molecule, thus, microsatellites are linked to that specific gene. See **microsatellite marker**.

- **microsatellite marker**

microsatellites or simple sequence repeats are a type of molecular markers, microsatellites consist of tandem repeats of 1-6 nucleotide motifs, the repeats usually are in units of ten or more, although repeats as small as six units have been found, the repeats can be (1) perfect tandem repeats, (2) imperfect (interrupted by several non-repeat nucleotides) or (3) compound repeats, they are well-distributed throughout a genome, microsatellites can be amplified by the polymerase chain reaction (PCR) using a pair of primers flanking the repeat sequence, the polymorphism between different individuals is due to the variation in the number of repeat units, each locus can have many alleles, one advantage of microsatellites is that they are mostly codominant, which make them easily transferable between genetic maps of different crosses in the same or closely related species, in contrast with RAPDs, which are dominant and therefore new maps have to be generated for every cross, several microsatellite-primer pairs may be used simultaneously, thus reducing time and costs, the relatively simple interpretation and genetic analysis of single-locus markers make them superior to multi-locus DNA marker types such as RAPDs, microsatellites are also called 'simple sequence repeats' (SSRs), 'simple tandem repeats' (STRs) or 'simple sequences' (SSs).

- **Microsatellite-Primed PCR (MP-PCR)**

a technique resulting in RAPD-like patterns after agarose gel electrophoresis and ethidium bromide staining, the MP-PCR technique is more reproducible than RAPD analysis because of higher stringency random amplified polymorphic (RAPD) technique.

- **microscope**
  an optical instrument having a magnifying lens or a combination of lenses for inspecting objects too small to be seen distinctly by the unaided eye.

- **microscopic slide**
  glass slide.

- **microsome**
  minichromosome.

- **microsporangium**
  a sporangium (e.g., pollen sac) that produces the microspores (pollen).

- **microspore**
  the first cell of the male gametophyte generation of Angiospermae and Gymnospermae, later to form the pollen grain.

- **microspore culture**
  it refers to the in vitro culture of pollen grains to obtain haploid callus or haploid plantlets directly from the pollen grains, microspore cultures differ from pollen cultures by the stage of development in gametogenesis.

- **microspore mother cell**
  one of the many cells in the microsporangium (anther) that undergo microsporogenesis to yield four microspores, as opposed to megaspore mother cell.

- **microsporocyte**
  Pollen Mother Cell (PMC).

- **microsporogenesis**
  the development of microspores from the microspore mother cell meiosis.

- **microstylous**
  showing a short stamen.

- **microsurgery**
  micromanipulation.

- **microsynteny**
  genomic relationships at gene level synteny.

- **microtome**
  a machine for cutting thin slices of embedded tissue, these sections may be stained and examined with the light or electron microscope.

- **microtubule**
  a tubular structure, 15-25 nm in diameter, of indefinite length and composed of subunits of the protein tubulin, it occurs in large numbers in all eukaryotic cells, either freely in the cytoplasm or as a structural component of organelles. They form part of the structure of the mitotic spindle, which is responsible for the movement of chromosomes during cell division.

- **microtubules**
  hollow cylinders made of the protein tubulin that form, among other things, the spindle fibres.

- **micrurgy**
  micromanipulation.

- **mictic**
  amphimictic.

- **micton**
  it refers to a species of wide distribution, which is the result of hybridisation of two or more species, all individuals are cross-fertile and have ancestral genotypes, apomixis is not present.

- **mid rip**
  the central, thick, linear structure that runs along the length of a plant lamina, it occurs in true leaves as a vein running from the leaf base to the apex. It provides support and is a translocative vessel.

- **midparent value**
  the mean of the values of a quantitative phenotype for two specific parents.

- **migration**
  movement of individuals between otherwise reproductively isolated populations.

- **milky stage**
  in cereals, the stage of 'milk' ripening of caryopses at which the endosperm shows a milky consistency.

- **milling**
  the processes in which cereal grains are subjected to grinding followed by sifting, sizing or other separation techniques, for example, in wheat, the grain is tempered to approximately 15-17 percent moisture, which facilitates the separation of the endosperm, the pericarp and embryo.

- **millipore filter**
  a disc-shaped synthetic filter having holes of specified diameter (0.005-8 μ) through its surface.

- **mimicry**
  a phenomenon in which an individual, gains an advantage by looking like the individuals of a different species.

- **mineral soil**
  a soil containing <20 percent organic matter or having a surface organic layer <30 cm thick.

- **mineralisation**
  the conversion of organic tissue to an inorganic state as a result of decomposition by the organic content.

- **minichromosome**
  a very small chromosome, usually as a result of chromosome aberrations.

- **minimal medium**
a culture medium for microorganisms that contains the minimal necessities for growth of the wild-type. A medium containing only inorganic salts, a carbon source and water.

- **miniprep**
an abbreviation for minipreparation, it refers to a small-scale preparation of plasmid or phage DNA commonly used after cloning to analyse the DNA sequence inserted into a cloning vector.

- **minor crops**
crops that may be high in value but that are not widely grown (e.g., many fruits, vegetables and trees.)

- **minor element**
trace element.

- **minor gene**
a gene that individually exerts a slight effect on the phenotype modifying (modifier) gene.

- **minor oilseeds**
oilseed crops other than soybeans and peanuts (e.g., in some countries, sunflower seed, canola, rapeseed, safflower, mustard seed and flax seed).

- **minute fragment (of a chromosome)**
usually very tiny chromosome segments as a result of chromosome aberrations.

- **misdivision**
aberrant chromosome division in which no longitudinal but transversal separation of the centromere occurs, the consequence may be telocentric chromosomes.

- **mismatch repair**
a form of excision repair initiated at the sites of mismatched bases in DNA.

- **mismatching**
a region of DNA in a heteroduplex where bases cannot pair mismatch repair.

- **mispairing**
the presence in one chain of a DNA double helix of a nucleotide that is not complementary to the nucleotide occupying the corresponding position in the other chain.

- **missense mutation**
mutations that at change a codon for one amino acid into a codon for a different amino acid.

- **mitochondrial cytopathy**
human disorder caused by point mutation or deletion in

mitochondrial DNA, inherited maternally.

■ **mitochondrial DNA**
the mitochondrial genome consists of a circular DNA duplex, with 5 to 10 copies per organelle.

■ **mitochondrion**
an oval, round or thread-shaped organelle, whose length averages 2 μm and that occurs in large numbers in the cytoplasm of eukaryotic cells. It is a double-membrane-bound structure in which the inner membrane is thrown into folds (cristae) that penetrate the inner matrix to varying depths. It is a semi-autonomous organelle containing its own DNA and ribosomes and reproducing by binary fission. It is the major site of ATP production and thus of oxygen consumption in cells.

■ **mitomycin c**
a form of a family of antibiotics produced by *Streptomyces caespitosus,* it prevents DNA replication by crosslinking the complementary strands of the DNA double helix.

■ **mitosis**
the process of nuclear division by which two daughter nuclei are produced, each identical to the parent nucleus before mitosis begins each chromosome replicates to two sister chromatids, these then separate during mitosis so that one duplicate goes into each daughter nucleus. In contrast to the prophase of meiosis I, the prophase of mitosis does not involve pairing of chromosomes or crossing-over between the homologous chromosomes. During metaphase, the individual chromosomes line up at the equatorial plate of the cell and a spindle fibre develops that links each of their chromatids to one of the two poles in the cell. The chromatids of each chromosome separate and move to opposite poles at anaphase. At telophase, a nuclear membrane develops

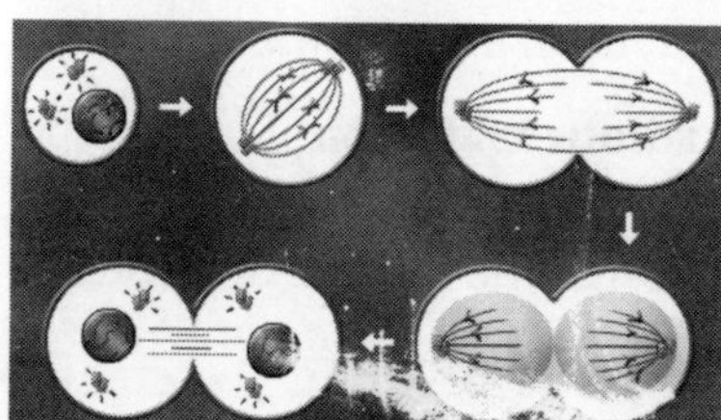

around the chromosomes to define the nucleus of the cell, a cell wall is formed.

■ **mitotic apparatus**
an organelle consisting of three components:

1. the asters, which form the centrosome
2. the gelatinous spindle

3. the traction fibres, which connect the centromeres of the various chromosomes to either centrosome

- **mitotic crossover**
a crossover resulting from the pairing of homologous chromosomes in a diploid during mitosis.

- **mitotic cycle**
the sequence of steps by which the genetic material is equally divided before the cell division into two daughter cells happens.

- **mitotic index (MI)**
the fraction of cells undergoing mitosis in a given sample, usually the fraction of a total of 1,000 cells that are undergoing division at one time.

- **mitotic inhibition**
induced or spontaneous inhibition of the mitotic division.

- **mitotic poison**
any substance that hampers the proper mitosis.

- **mitotic recombination**
the recombination of genetic material during mitosis and the process of asexual reproduction, the mechanism for the production of variation in heterokaryons.

- **mitotic spindle**
the spindle-shaped system of microtubules that, during cell division, traverses the nuclear region of eukaryotic cells, the chromosomes become attached to it and it separates them into two sets, each of which can be enclosed in the envelope of a separate daughter nucleus.

- **mixed codon family**
group of four codons sharing their first two bases but coding for more than one amino acid.

- **mixoploid**
cell populations in which different cells show different chromosome numbers mosaicism.

- **mixture**
it consists of seed of more than one kind or variety, each present in excess of at least 5% of the whole blend.

- **modal value**
mode.

- **mode**
the single class in a statistical distribution having the greatest frequency.

- **mode of reproduction**
two general modes of reproduction are distinguished: (1) sexual and (2) asexual. Sexual reproduction involves the union of

male and female gametes derived from the same or different parents. Asexual reproduction occurs by multiplication of plant parts or by seed production that does not involve the union of sexual gametes, the breeding procedures are dependent on the mode of reproduction.

- **modern varieties**
  varieties developed by breeders in the formal system.
- **modificability**
  the ability of phenotypic variation of a particular genotype in response to varying environmental conditions.
- **modification**
  non-heri morphological or physiological changes induced by varying abiotic or biotic influences. In molecular biology, modification of DNA by DNA methylases occurs after replication, site-specific methylation protects the DNA (e.g., of bacteria), which synthesise restriction endonucleases.
- **modifier gene**
  a gene that affects the phenotypic expression of another gene.
- **moisture tension**
  the force at which water is held by soil. It is expressed as the equivalent of an unit column of water in centimetres.
- **molasses**
  residue of the beet sugar or sugarcane production.
- **mouldboard**
  the curved metal plate in a plough that turns over the earth from the furrow.
- **mouldboard plough**
  a plough with a point and a heavy curved blade for breaking the soil.
- **molecular chaperone**
  a protein that aids in the folding of a second protein. The chaperone prevents proteins from taking conformations that would be inactive.
- **molecular cloning**
  DNA segments of different sises of prokaryotic or eukaryotic origin are identically multiplied as a part of bacterial plasmids or phages: the alien DNA is incorporated into the host DNA ring molecule, by the rapid division of the host cells the alien DNA segments are simultaneously cloned, subsequently, those cloned DNA segments can be excised, separated and purified for further utilisation.
- **molecular genetics**
  a branch of genetics that deals with molecular aspects of genetic mechanisms.

- **molecular hybridisation**
  the annealing of previously purified and denatured DNA strands by different means DNA hybridisation.

- **molecular imprinting**
  the phenomenon in which there is differential expression of a gene depending on whether it was maternally or paternally inherited. Paternal imprinting means that an allele inherited from the father is not expressed in offspring. Maternal imprinting means that an allele inherited from the mother is not expressed in offspring.

- **molecular marker**
  particular DNA sequences and/or segments that are closely linked to a gene locus and/or a morphological or other characters of a plant. Those segments can be detected and visualised by molecular techniques. Roughly, three groups of markers can be classified:
  1. hybridisation-based DNA markers such as Restriction Fragment Length Polymorphisms (RFLPs) and oligonucleotide fingerprinting,
  2. PCR-based DNA markers such as Random Amplified Polymorphic DNAs (RAPDs), which can also be converted into Sequence Characterised Amplified Regions (SCARs), Simple Sequence Repeats (SSRs) or microsatellites, Sequence-Tagged Sites (STS), Amplified Fragment Length Polymorphisms (AFLPs), Inter-Simple Sequence Repeat Amplification (ISA), Cleaved Amplified Polymorphic Sequences (CAPs) and Amplicon Length Polymorphisms (ALPs),
  3. DNA chip and sequencing-based DNA markers such as single nucleotide polymorphisms (SNPs).

- **molecular weight (MW)**
  the sum of the atomic weights of all of the atoms in a given molecule.

- **molecule**
  that ultimate unit quantity of a compound that exists by itself and retains all the chemical properties of the compound.

- **molybdenum (Mo)**
  an element that is required in small amounts by plants and is found largely in the enzyme nitrate reductase, deficiency leads to interveinal chlorosis.

- **monad**
  a meiocyte-derived individual cell instead of a tetrad as a result of meiotic disturbances.

- **monoallelic**
applied to a polyploid in which all alleles at a particular locus are the same (in a tetraploid—*A1A1A1A1*), as opposed to diallelic (in a tetraploid—*A1A1A1A2*), triallelic (in a tetraploid—*A1A1A2A3*), tetraallelic (in a tetraploid—*A1A2A3A4*), etc.

- **monobrachial**
a chromosome with a terminal centromere telocentric.

- **monocarpic**
bearing one fruit.

- **monocentric chromosome**
a chromosome with only one centromere 11 neocentric polycentric.

- **monocentric crop plant**
a crop plant species that has only one centre of origin centre of diversity.

- **monocistronic mRNA**
an mRNA that codes for only one protein.

- **monoclonal**
describing genetically identical cells produced from one clone gene.

- **monoclonal blocks**
a deployment option for clones or for families in monofamily blocks in which each clone (family) is established in a pure block, diversity may be maintained by a mosaic of blocks of different genetic entries.

- **monocot**
an abbreviated name for monocotyledon, referring to plants having single-seed leaves, flower parts arranged in threes or multiples thereof, parallel-veined leaves, closed vascular bundles arranged randomly in the stem tissue.

- **monocotyledonous**
having one cotyledon monocot.

- **monoculm mutant**
a mutant form that shows only one culm instead of normally more (e.g., in wheat).

- **monoculture**
the growing over a large area of a single crop species or of a single variety of a particular species.

- **monoecious plant**
a plant species in which male and female organs are found on the same plant but in different flowers (for example maize).

- **monoecism**
the condition in plants that have male and female flowers separated on the same plant (e.g., in maize or pumpkin.)

- **monogenic**
  a trait controlled by the alleles of one particular locus, as opposed to digenic, trigenic oligogenic or polygenic.
- **monogenic resistance**
  resistance determined by a single gene.
- **monogenomatic**
  monohaploid.
- **monogenotypic**
  clone variety.
- **monogerm(ous)**
  a fruit of, for example, sugarbeet containing only one ovule in contrast to a multigerm fruit, which represents an aggregate fruit containing several ovule units.
- **monohaploid**
  a haploid cell or individual possessing only one chromosome set in the nucleus.
- **monohybrid**
  1. offspring of parents that differ in only one genetic characteristic. Usually implies heterozygosity at a single locus under study.
  2. a cross between two individuals that are identically heterozygous for the alleles of one particular gene (i.e., $Aa \times Aa$).
- **monohybrid heterosis**
  superdominance.
- **monohybrid segregation**
  a segregation pattern according to a monogenic inheritance.
- **monoisodisomic**
  a cell or individual showing monosomy for one chromosome but an isochromosome for one of the arms of the missing chromosome.
- **monoisosomic**
  a cell or individual showing nullisomy of one chromosome but having an isochromosome for one arm of the missing chromosome pair.
- **monophyletic**
  a group of species that share a common ancestry, being derived from a single interbreeding population.
- **monophyllous**
  showing one leaf.
- **monoploid**
  a cell having only one chromosome set (usually as an aberration) or an organism composed of such cells.
- **monoploid**
  having the basic chromosome number in a polyploid series.
- **monoplontic**
  it refers to a haploid individual or monoploid phase of the life cycle.

- **monosome**

a chromosome that lacks a homologue in a diploid organism.

- **monosomic**

a diploid cell missing a single chromosome. A cell or individual that is basically diploid but that has only one copy of one particular chromosome type and thus has chromosome number 2n - 1.

- **monosomic analysis**

a common method for gene mapping (e.g., in hexaploid wheat), when genes determining phenotypes of interest for which an aneuploid series is not available, crosses can be made to a monosomic series in a variety with a contrasting phenotypic pattern. The monosomics are used as female parents in order to ensure the majority of progeny (~ 72 percent, the transmission of $n - 1$ gametes is about 75 percent by the egg cell, but only about 4 percent by the pollen) will be become monosomic, if the gene involved is both recessive and hemizygous-effective (an uncommon situation), direct phenotypic observations on F1 monosomic progenies enable the researcher to locate the gene on a particular chromosome, only monosomic individuals in the critical cross will exhibit the phenotype, whereas disomic sibs and monosomic (or disomic) progenies from all other crosses will display the dominant phenotype. If the plants of the monosomic series carry the dominant allele of a gene of interest, monosomic individuals in the critical cross exhibit the recessive phenotype, whereas disomic sibs display the dominant phenotype aneuploid.

- **monospermic**

monospermous.

- **monospermous**

bearing one seed.

- **monostand**

sometimes it refers to a grass community composed of only one cultivar.

- **monotelic**

a mitotic chromosome with one oriented and one unoriented centromere centromere.

- **monotelocentric**

a cell or individual lacking one chromosome pair but showing one telocentric chromosome for one arm of the two missing homologues.

- **monotelodisomic**

a cell or individual lacking one chromosome pair but showing two homologous telocentric chromosomes for one arm of the two missing homologues.

- **monotelomonoisosomic**
  a cell or individual lacking one chromosome pair but showing a telocentric chromosome for one arm of the missing homologous pair and an isochromosome for the other arm.

- **monotelotrisomic**
  a cell or individual showing an additional telocentric chromosome to a certain pair of chromosomes.

- **mordant**
  mordanting.

- **mordanting**
  in a broad sense, to produce surface conditions by metal ions in the fixed structures that will enable them to hold the particular stains intended for making them visible.

- **Morgan unit, Morgan (M)**
  a unit of relative distance between genes on a chromosome, one Morgan (1 M) represents a crossing-over value of 100 percent, a crossing-over value of 10 percent is a decimorgan (dM), 1 percent is a centimorgan (cM).

- **morphogenesis**
  the developmental processes leading to the characteristic mature form of a plant or parts of it.

- **morphological species concept**
  organisms are classified in the same species if they appear identical by morphological (anatomical) criteria.

- **morphosis**
  a modification of the morphogenesis of an individual caused by environmental changes.

- **morphotype**
  habit(us).

- **mosaic**
  a chimera, a tissue containing two or more genetically distinct cell types or an individual composed of such tissues. Individual made up of two or more genetically distinct cell lines.

- **mosaicism**
  intraindividual variation of chromosome numbers or chromosome structure, usually in different tissues.

- **mother cell**
  special cells in the anther and ovule that give rise to pollen or egg cells.

- **mother plant**
  the female ancestor of a hybrid and/or hybrid progeny, in horticulture, a mature plant from which cuttings are taken donor plant.

▪ **motif**
a characteristic stretch of amino acids in a protein sequence which is responsible for a biochemical function, e.g., the active site of an enzyme.

▪ **mound layering**
a method of propagation whereby a branch or stem is scored and then brought into contact with the soil to spur rooting.

▪ **MP-PCR**
Microsatellite-Primed PCR.

▪ **mRNA**
messenger RNA, an RNA molecular that functions during translation to specify the sequence of amino acids in a nascent polypeptide.

▪ **mtDNA**
mitochondrial DNA.

▪ **mu genes**
a class of genes that frequently spontaneously mutate.

▪ **mu particle**
bacteria-like particle found in the cytoplasm of Paramecium that have the mate-killer phenotype.

▪ **mu phage**
a kind (species) of phage with properties similar to those of insertion sequences, being able to insert, transpose and cause chromosome rearrangements and mutations.

▪ **MU site**
mutational site.

▪ **mucilage**
the gummy, sticky complex carbohydrate (consisting principally of polyuronides and galacturonides that chemically resemble the pectic compounds and hemicellulose) substances that cover the seeds, the root tip, bark or stems of some plants.

▪ **mugeinic acid**
a chelating agent, it plays an important role in the uptake of heavy metal ions from the soil when it is exudated by roots of some gramineous plants (e.g., of rye) chelate.

▪ **mule**
a plant hybrid that is self-sterile and usually cross-sterile due to infertile pollen or undeveloped pistils.

▪ **mullerian mimicry**
a form of mimicry in which noxious species evolve to resemble each other.

▪ **multidimensional scaling**
a multivariate method of showing similarities and/or dissimilarities of empirical data (e.g., genotypes) within an n -dimen-

sional euclidic space (e.g., $n = 2$ or 3) in which the distances between the objects are as best as possible arranged according to the distances of the data matrix.

■ **multifactorial**
a characteristic influenced in its expression by many factors, both genetic and environmental.

■ **multigene family**
multigene variety.

■ **multigene variety**
a variety that carries a number of specific genes governing resistance to a particular pathogen.

■ **multigenic**
a trait that is controlled by many genes, as opposed to monogenic polygenic.

■ **multigerm**
an aggregate fruit containing several ovules monogerm(ous).

■ **multihybrid**
1. an organism heterozygous at numerous loci.
2. an individual that is heterozygous for more than one gene.

■ **multiline**
a cultivar or variety that is composed by many more or less defined lines multiline variety.

■ **multiline variety**
a composite (blended) population of several genetically related lines of a self-pollinated crop, but bearing different genes (e.g., for resistance to pathogens.)

■ **multilineal variety**
multiline variety.

■ **multilocation testing**
a testing of breeder's strains and varieties on several geographically different sites in order to estimate the adaptive environmental response and/or performance stability.

■ **multi-locus probe**
a DNA probe that hybridises to a number of different sites in the genome of an organism.

■ **multimeric structure**
a structure composed of several identical or different subunits held together by weak bonds.

■ **multinomial expansion**
the terms generated when a multinomial is raised to a power.

■ **multinucleate**
describes cells that have more than one nucleus.

■ **multiple alleles**
the existence of several known allelic forms of a gene allelism.

■ **multiple allelism**
the existence of several known alleles of a gene.

- **multiple cropping**
  the growing of more than one crop on the same field in one year.

- **multiple fruit**
  developed from a cluster of flowers on a common base.

- **multiple genes**
  two or more genes at different loci that produce complementary or cumulative effects on a single, quantitative genetic trait polygenes.

- **multiple population breeding system**
  the breeding population is subdivided in several smaller subpopulations that are bred for different objectives, e.g., different areas.

- **multiple-factor hypothesis**
  a hypothesis to explain quantitative variation by assuming the interaction of a large number of genes (polygenes) each with a small additive effect on the character.

- **multiple-hit hypothesis**
  the proposal that a single cell must receive a series of mutational events in order to become malignant or cancerous.

- **multiplexing**
  a sequencing approach that uses several pooled samples simultaneously, greatly increasing sequencing speed.

- **multiplication**
  the increase in number of individuals produced from seed or by vegetative means.

- **multiplication population**
  subpopulations of the general breeding population purposefully selected for different sets of traits or deployment destinations.

- **Multiplicity Of Infection (MOI)**
  the average number of phage particles that infect a single bacterial cell in a specific experiment.

- **multiply**
  multiplication.

- **multitude of genes**
  polygenes.

- **multivalent**
  designating and association of more than two chromosomes whose homologous regions are synapsed by pairs (e.g., in autopolyploids or in translocation heterozygotes.)

- **multivar**
  cultivar mixture.

- **mummy**
  a dried, shrivelled fruit or seed colonised by a fungus or parasite.

- **mutability**
  the ability of a gene to undergo mutation.

- **mutagen**
  an agent that increases the mutation rate within an organism or cell, for example, X-rays, gamma-rays, neutrons or chemicals (base analogues, such as 5-bromo uracil, 5-bromo deoxyuridine, 2-amino purine, 8-ethoxy caffeine, 1.3.7.9.-tetramethyl-uric acid, maleic hydrazide, antibiotics, such as azaserine, mitomycin C, streptomycin, streptonigrin, actinomycin D, alkylating agents, such as sulfur mustards [ethyl-2-chloroethyl sulfide], nitrogen mustards [2-chloroethyl-dimethyl amine], epoxides [ethylene oxide], ethyleneimines, sulfates, sulfonates, diazoalkanes, nitroso compounds [*N*-ethyl-*N*-nitroso urea], azide [sodium azide], hydroxylamine, nitrous acid, acridines [hydrocyclic dyes], such as acridine orange.

- **mutagenesis**
  the process leading to a mutant genotype.

- **mutagenic**
  substances and circumstances inducing mutants.

- **mutagenic agent**
  mutagen.

- **mutagenicity**
  the potential of agents and circumstances to induce mutations.

- **mutagenicity testing**
  the assessment of chemical or physical agents for mutagenicity.

- **mutagenise**
  treatments that result in mutations.

- **mutagenised**
  cells or individuals that were treated with mutagens.

- **mutant**
  a plant bearing a mutant gene that expresses itself in the phenotype.

- **mutant allele**
  an allele differing from the allele found in the standard or wild type organism.

- **mutant screening**
  the process of screening for mutations showing abnormalities in a certain structure or in a particular function, as a preparation for genetic analysis of that structure or function.

- **mutant site**
  a site on a chromosome at which a mutation can occur or has occurred.

■ **mutant strain**
a strain of cells or individuals that, by one or more mutations, is differentiated from the original strain.

■ **mutation**
1. the process producing a gene or a chromosome differing from the wild-type.
2. The gene or chromosome that results from such a process.

■ **mutation breeding**
to experimentally introduce or remove a character from a cell or organism by exposure to mutagenic agents followed by screening for the desired attribute. It also refers to several techniques involving induced mutations that were utilised (mainly in the 1960s and 1970s) to introduce desirable genes into the plants (e.g., resistance to plant diseases, increased yield, improvements in composition). Usually seeds or pollen were soaked in mutation-causing chemicals (mutagens) or via bombardment treated with ionising radiation followed by screening of the resultant plants and selection of the particular mutation (beneficial trait).

■ **mutation event**
the actual occurrence of a mutation in time and space.

■ **mutation frequency**
the frequency of mutations in a population.

■ **mutation map**
the frequency of mutations recorded along of chromosomes, represented as a diagrammatic drawing.

■ **mutation pressure**
the continued production of an allele by mutations.

■ **mutation rate**
the number of mutation events per gene and per unit of time (e.g., per cell generation.)

■ **mutational dissection**
the study of the components of a biological system through a study of mutations affecting that system.

■ **mutational hot spot**
a site within a gene or genome that frequently mutates.

■ **mutational site**
the more or less defined position along a gene at which mutations occur.

■ **mutator gene**
a gene that may increase the spontaneous mutation rate.

■ **mutator mutation**
mutation of DNA polymerase that increases the overall mutation rate.

muton
a term coined by Benzer for the smallest mutable site within a cistron. The smallest part of a gene that can be involved in a mutation event, now known to be a single nucleotide pair.

mycelium
a mass of hyphae that form the body of a fungus.

mycobacterium
any of several rod-shaped aerobic bacteria of the genus *Mycobacterium*.

mycology
the science of fungi, the study of mushrooms.

mycor(r)hiza
a close physical association between a fungus and the roots (or seedlings) of a plant from which both fungus and plant appear to benefit.

mycotoxin(s)
toxic substances produced by fungi or moulds on agricultural crops.

myotonic dystrophy
a combination of progressive weakening of the muscles and muscle spasms or rigidity, with difficulty relaxing a contracted muscle, inherited as an autosomal dominant trait.

n
the symbol for the haploid chromosome number.

n banding
a special chromosome staining method related to C banding, which reveals specific types of heterochromatin, the pattern of bands and interbands along a chromosome serves as a tool for chromosome or chromosome segment identification. The method was successfully applied in wheat C banding.

n segment
sequence of nucleotides added in a template-independent fashion at the joining junctions of heavy-chain immunoglobulin genes.

naked (barley or oats)
mutant varieties that thresh free from their husk in contrast to conventional varieties where the husk is held firmly to the grain.

nanometer
equals 10^{-9} meter equals 10 Ångström.

nanovirus
geminivirus.

nap
large masses of curled and loosely matted fibres found in raw cotton.

- **Naphthalene Acetic Acid (NAA)**
  a synthetic auxin.

- **napiform**
  turniplike in form.

- **narrow wing**
  the smaller of the two parts of, for example, the wheat glume, which are separated by the keel.

- **narrow-leafed**
  angustifoliate.

- **nastic movement**
  nasty.

- **nasty**
  the response of a plant organ to a nondirectional stimulus (e.g., light), it is facilitated by changes in cell growth or changes in turgor.

- **native**
  indigenous.

- **native breed**
  landrace.

- **native DNA**
  double-stranded DNA isolated from a cell with its hydrogen bonds between strands intact, as opposed to denatured DNA.

- **natural selection**
  a complex process in which the total environment determines which members of a species survive to reproduce and so pass on their genes to the next generation.

- **nature reserve**
  an area of land set aside for nature conservation and associated scientific research, usually with strong legal protection against other uses.

- **nearest-neighbour analysis**
  a technique of transferring radioactive atoms between adjacent nucleotides in DNA used to demonstrate that the two strands of DNA run in opposite directions.

- **near-isogenic lines**
  not fully isogenic, for example, in maize, two distinct composites of F3 lines from a single cross, one consisting of lines homozygous recessive and the other consisting of lines homozygous dominant for a certain gene (i.e., there is same genetic background), however, differing only in being homozygous dominant versus recessive for the genes, in wheat, near-isogenic lines were produced for different *Rht* (reduced height) genes causing different straw length.

- **neck**
  the uppermost part of the culm between the flagleaf sheath and the collar.

- **necrosis**
  death of plant tissue, which is usually accompanied by discol-

oration or becoming dark in colour, commonly a symptom of fungus infection.

- **necrotrophic pathogen(e)**
  a fungal pathogen that causes the immediate death of the host cells as it passes through them, a coloniser of dead tissue.

- **nectar**
  a sticky, sometimes sweet, secretion of flowers, which has an attraction for insects.

- **nectar gland**
  nectarium nectary.

- **nectarine**
  a variety of peach having a smooth, downless skin peach.

- **nectarium**
  nectary.

- **nectary**
  refers to a sugar-secreting gland, nectaries are usually situated at the base of a flower, sometimes in a spur, in order to attract pollinators. Nectaries can also be extrafloral (e.g., the gland spin of certain cacti where they attract seed dispersal insects, such as ants).

- **needle**
  a linear, commonly pungent leaf.

- **negative assortative mating**
  preferential mating between phenotypically different partners.

- **negative control**
  transcription regulation mediated by factors that block, turn down or turn off transcription.

- **negative interference**
  the phenomenon whereby a crossover in a particular region enhances the occurrence of other apparent crossovers in the same region of the chromosome.

- **neighbourhood**
  a partially isolated subpopulation with a certain degree of inbreeding, it may arise when a large population splits into subpopulations by inbreeding effects.

- **nema**
  eelworm.

- **nematicid**
  a substance or preparation used for killing nematodes parasitic to plants.

- **nematode**
  a microscopic soil worm that may attack roots or other structures of cereal, sugarbeet, potato and other plants and cause extensive damage.

- **n-end rule**
the life span of a protein is determined by its amino-terminal (N-terminal) amino acid.

- **neocentric**
secondary centromeres, that under certain conditions, show movement as the primary centromere. Sometimes they are observed on chromosome ends when they move toward the poles during anaphase of meiosis I.

- **neocentric activity**
neocentric.

- **neo-Darwinian evolution**
evolutionary theory incorporating Darwinism and Mendeliań genetics.

- **neo-Darwinism**
the merger of classical Darwinian evolution with population genetics.

- **neomorph**
mutant showing a novel substance or structure not found in wild-types.

- **neoplasm**
new growth of abnormal tissue.

- **nervation (of leaf)**
the arrangement of veins in a leaf.

- **nervature**
nervation.

- **nerve**
the line, usually raised, on the surface of a lemma or glume marking the presence of conducting tissue below the surface.

- **nervure**
leaf vein nervation.

- **nest planting**
setting out a number of seedlings or seeds close together in a prepared hole, pit or spot.

- **netting**
covering plots or individual plants with nets in order to prevent bird damage.

- **neurospora**
a pink mould, commonly found growing on old food.

- **neutral gene hypothesis**
the hypothesis that most genetic variation in natural populations is not maintained by selection because most alleles have equal fitness.

- **neutral mutation**
1. a mutation that has no effect on the Darwinian fitness of its carriers.
2. a mutation that has no phenotypic effect.

- **neutral petite**
a petite yeast that produces all wild-type progeny when crossed with wild-type.

■ nick

the two parents for producing hybrid seed when they produce high yields of seed of a highly productive and desirable hybrid. In breeding, synchronisation of the receptivity of the female organ to the maximum pollen load of the pollinator for cross-fertilisation. In molecular genetics, a single-strand break of DNA nick translation.

■ nick translation

a technique by which a DNA molecule is radioactively labelled with high specificity, such labelled DNA is used by different DNA hybridisation methods as a probe (e.g., Southern transfer), within a double-stranded DNA several single-strand breaks (nicks) are produced by DNase hydrolysis, each break or gap is extended by 5'>3' exonuclease activity of DNA polymerase I, the removed 5' nucleotides are immediately substituted by the polymerase activity of the same exonuclease. However, now ^{32}P-deoxynucleotides are used, which label the newly synthesised DNA.

■ nicking

nuclease action to sever the sugar-phosphate backbone in one DNA strand but not the other at one specific site.

■ nicotianamide

a soluble crystal amide of nicotinic acid that is a component of the vitamin B complex.

■ nicotine

a colourless, oily, water-soluble, highly toxic liquid alkaloid, $C_{10}H_{14}N_2$, found in tobacco and valued as an insecticide biological control.

■ nif genes

the genetic designation of genes participating in the process of nitrogen fixation. About 17 genes are organised in the *nif* operon, by the *nif* genes produced proteins the atmospheric nitrogen (N_2) will be fixed as NH_4^+ and NO_3^- ions, many soil bacteria may fix atmospheric nitrogen. There are many research activities dealing with the transfer of the bacterial system of nitrogen fixation into crop plants other than legumes.

■ nitrification

the oxidation of ammonia to nitrite and nitrite to nitrate by chemolithotrophic bacteria.

■ nitrocellulose

a nitrated derivative of cellulose, it is used in the form of a membrane as a filter for macromolecules in blotting techniques.

- **nitrocellulose filter**
a type of filter used to bind DNA for hybridisation.

- **nitrogen (N) consumers**
a crop plant in the crop rotation that takes nitrogen up from soil, as opposed to plants (e.g., legumes) that provide nitrogen to the soil by their nitrogen fixation activity nitrogen fixation.

- **nitrogen (N) fixation**
the reduction of gaseous molecular nitrogen and its incorporation into nitrogenous compounds, it is facilitated by lighting, photochemical fixation in the atmosphere and by the action of nitrogen-fixing microorganisms (bacteria).

- **nitrogen (N)**
an element that is essential to all plants, it is found reduced and covalently bound in many organic compounds and its chemical properties are especially important in the structure of proteins and nucleic acids, deficiency causes chlorosis and etiolation.

- **nitrogen base**
type of molecule that forms an important part of nucleic acid, composed of a nitrogen-containing ring structure. Hydrogen bonds between bases in opposing complementary strands link the two strands of a DNA double helix.

- **nitrogenase**
an enzyme complex that catalyses the reduction of molecular nitrogen in the nitrogen-fixation process in which dinitrogen is reduced to ammonia.

- **nobilisation**
a term used in the breeding of sugarcane to indicate repeated matings (backcrossing) to the 'noble' canes (i.e., restoring intergeneric *Saccharum* hybrids to the phenotype of *Saccharum officinarum*).

- **nodal bud**
the lateral shoot bud located within the root ring at the node.

- **node**
a slightly enlarged portion of a stem where leaves and buds arise and where branches originate.

- **nodulation**
in legumes, species of *Rhizobium* bacteria fixing nitrogen of the air in association with the roots on which they are the cause of swellings (nodules).

- **nodule**
a small, hard lump or swelling, root nodules are characteristic of *Rhizobium* infection and nitrogen fixation in legumes.

- **nodule bacteria**
  it refers to several species of nitrogen-fixing *Rhizobium* bacteria, which form ball-like nodules along legume roots.

- **nodus**
  node.

- **nominal scale**
  a scale for scoring quantitative data using a series of predefined values (e.g., flower colour).

- **noncoding DNA**
  a certain portion of DNA that obviously does not determine a gene product, such as a protein and/or character.

- **non-coding strand**
  see **anticoding strand.**

- **nonconjunction**
  the failure of metaphase chromosome pairing during meiosis.

- **nondemanding variety**
  low-input variety.

- **nondisjunction**
  the failure of separation of paired chromosomes at metaphase, resulting in one daughter cell receiving both and the other daughter cell none of the chromosomes in question. It can occur both in meiosis and mitosis.

- **non-disjunction**
  the failure of a pair of homologous chromosomes to separate properly during meiosis. The failure of homologues (at meiosis) or sister chromatids (at mitosis) to separate properly to opposite poles, that is two chromosomes or chromatids go to one pole and none to the other.

- **non-histone protein**
  the protein remaining in chromatin after the histones are removed. The scaffold structure is made of nonhistone proteins.

- **nonhomologous association**
  pairing of chromosomes, which obviously are not homologous, however, it is presumed that cryptically homologous segments allow the chromosome association.

- **nonhomologous chromosomes**
  the different chromosomes of a haploid chromosome set, which usually cannot pair with another.

- **nonhost resistance**
  inability of a pathogen to infect a plant because the plant is not a host of the pathogen due to lack of something in the plant that the pathogen needs or to the presence of substances incompatible with the pathogen.

▪ **nonimmune**
susceptible.

▪ **noninfectious disease**
a disease that is caused by an abiotic agent (i.e., by an environmental factor, not by a pathogen.)

▪ **non-linear tetrad**
a tetrad in which the meiotic products are in no particular order.

▪ **non-Mendelian inheritance**
an unusual ratio of progeny phenotypes that does not reflect the simple operation of Mendel's law, for example, mutant: wild-type ratios of 3 : 5, 5 : 3, 6 : 2 or 2 : 6, indicate that gene conversion has occurred. In general, it refers to extrachromosomal and/or nonchromosomal inheritance.

▪ **nonparametric tests**
these are tests that do not make distributional assumptions, particularly the usual distributional assumptions of the normal-theory-based tests. Nonparametric tests usually drop the assumption that the data come from normally distributed populations.

▪ **non-parental**
see **recombinants**.

▪ **Non-Parental Ditype (NPD)**
a spore arrangement in Ascomycetes that contains only the two recombinant-type ascospores (assuming two segregating loci). A tetrad type containing two different genotypes, both of which are recombinant.

▪ **nonpreference**
a term used to describe a resistance mechanism where parasites prefer to be on some host genotypes more than others, the less preferred genotypes are resistant antixenosis.

▪ **nonrandom mating**
a mating system in which the frequencies of the various kinds of matings with respect to some trait or traits are different from those expected according to chance.

▪ **non-recombinant**
in mapping studies the offspring that have alleles arranged as in the original parents are non-recombinants.

▪ **nonrecurrent apomixis**
it refers to occasional apomixis, usually caused by haploid parthenogenesis.

▪ **nonrecurrent parent**
a parent that is not involved in a backcross.

- **nonsense codon**
one of the mRNA sequences (UAA, UAG, UGA) that signals the termination of translation. A codon for which no normal tRNA molecule exists. The presence of a nonsense codon causes termination of translation (ending polypeptide chain synthesis). There are three nonsense codons called amber (UAG) ochre(UAA) and opal (UGA).

- **nonsense mutation**
a mutation in which a codon is changed to a stop codon, resulting in a truncated protein product.

- **nonsense suppressor**
a mutation in the anticodon of tRNA that alters the anticodon so it is now complementary to a nonsense codon allowing the tRNA to insert its cognate amino acid at this nonsense codon during translation.

- **nonsister chromatid**
nonsibling chromatid.

- **nonspecific resistance**
horizontal resistance.

- **nontill rotation**
a method of planting crops that involves no seedbed preparation in the rotation other than opening small areas in the soil for placing seed at the intended depth, moreover, there is no cultivation during crop production, despite chemicals that are used for vegetation control.

- **nonuniform resistance**
resistance.

- **noonan syndrome**
a condition characterised by short stature and ovarian or testicular dysfunction, mental deficiency and lesions of the heart.

- **nopaline**
a rare derivate of an amino acid, it is produced in some crown galls of plants, the controlling genes are part of the t-sDNA of Ti plasmids.

- **norm of reaction**
the pattern of phenotypes produced by a given genotype under different environmental conditions.

- **normal curve**
normal distribution.

- **normal distribution**
1. any of a family of bell-shaped frequency curves whose relative position and shape are defined on the basis of the mean and standard deviation.
2. the most commonly used probability distribution in statistics, in nature, a vast number of continuous distributions are normally distributed.

- **normalising selection**
  the removal of genes and/or alleles that produce deviations from the normal phenotype of a population.

- **novel plant**
  intergeneric cross.

- **novel variety**
  it refers to seed, transplants or plants showing sufficient distinctness in the sense that the variety clearly differs by one or more identifiable morphological, physiological or other characteristics.

- **noxious**
  injurious (e.g., a noxious weed is one that crowds out desirable crops, robs them of plant food and moisture and causes extra labour in cultivation.)

- **noxious weeds**
  undesirable plants that infest either land or water resources and cause physical and economic damage noxious.

- **N-type**
  in sugarbeet breeding, varieties with normal sugar content and normal yielding capacity (N = Normal).

- **nucellar embryony**
  a way of parthenogenesis in which the embryo arises directly from the nucellus.

- **nucellus**
  the mass of tissue in the ovule of a plant that contains the embryo sac, size and shape can be diagnostic for species.

- **nuclear division**
  the division of the cell nucleus by mitosis, meiosis or amitosis.

- **nuclear division cycle**
  the sequence of stages of the division of the nucleus.

- **nuclear gene**
  a gene that is located on the chromosome of the nucleus.

- **nuclear pore**
  it allows the selective passage of materials into and out of the nucleus nuclear membrane.

- **nuclear staining**
  nuclear stain.

- **nuclear transplantation**
  1. the technique of placing a nucleus from another source into an enucleated cell.
  2. the transfer of a nucleus into the cytoplasm of another cell.

- **nuclease**
  one of the several classes of enzymes that degrade nucleic acid. An enzyme that can degrade DNA or RNA by breaking phosphodiester bonds.

- **nuclease hypersensitive site**
region of eukaryotic chromosome DNA that is specifically vulnerable to nuclease attack perhaps because it is not wrapped in histone as nucleosomes.

- **nucleic acid**
a large molecule composed of *nucleotide* subunits.

- **nucleoid**
a DNA mass within a chloroplast or mitochondrion.

- **nucleolar chromosome**
the chromosome that carries the nucleolus organiser (region), it may also be called a satellite chromosome.

- **nucleolar constriction**
that region of a chromosome that carries the nucleolus organiser. Besides the centromeric region it is observed as a secondary constriction along particular chromosomes. It is not stained by the standard chromosome techniques. By the nucleolar constriction the chromosome arm appears divided into two parts, the terminal part is called satellite, the whole chromosome is called the satellite chromosome.

- **nucleolar dominance**
amphiplasty.

- **nucleolar organiser**
the chromosomal region around which the nucleolus forms, a site of tandem repeats of the rRNA gene. A region (or regions) of the chromosome set physically associated with the nucleolus and containing rRNA genes.

- **nucleolar zone**
a chromosome region that is associated with the formation of the nucleolus during telophase nucleolar constriction.

- **nucleolin**
one of the nonribosomal proteins, it is considered to play a key role in regulation of rDNA transcription, perisomal synthesis, ribosomal assembly and maturation. It influences the nucleolar chromatin structure through its interaction with DNA and histones, it is involved in cytoplasmic-nucleolar transport of preribosomal particles ribonucleoprotein.

- **nucleolus**
a clearly defined, often spherical area of the eukaryotic nucleus, composed of densely packed fibrils and granules. Its composition is similar to that of chromatin, except that it is very rich in RNA and protein. It is the site of the synthesis of riboso-

mal RNA. The assembly of ribosomes starts in the nucleolus but is completed in the cytoplasm.

- **nucleolus organiser region (nor)**
nucleolar zone.

- **nucleoprotein**
a conjugated protein, composed of a histone or protamine bound to a nucleic acid as the nonprotein portion.

- **nucleosid(e)**
a glycoside that is composed of ribose or deoxyribose sugar bound to a purine or pyrimidine base.

- **nucleoside**
a sugar-base compound that is a nucleotide precursor. Nucleotides are nucleoside phosphates. A nitrogen base linked to a sugar molecule.

- **nucleosome**
a nu body, the basic unit of eukaryotic chromosome structure, a ball of eight histone molecules wrapped about by two coils of about 220 base pairs of DNA. Arrangement of DNA and histones forming regular spherical structures in eukaryotic chromatin.

- **nucleotide**
subunit that polymerises into nucleic acids (DNA or RNA). Each nucleotide consists of a nitrogenous base, a sugar and one to three phosphate groups.

- **nucleotide sequence**
the order of nucleotides along a DNA or RNA strand.

- **nucleotide synthesis**
See **biotechnology**.

- **nucleotide-pair substitution**
the replacement of a specific nucleotide pair by a different pair, often mutagenic.

- **nucleus**
the double-membrane-bound organelle containing the chromosomes that is found in most nondividing eukaryotic cells. It disappears temporarily during cell division. Within the nucleus several independent approaches point to the compartmentalisation of particular activities such as transcription, RNA processing and replication, chromosomes are revealed to occupy defined domains and to represent highly differentiated structures. The numerous activities that use DNA and RNA as a template occur with a defined spatial and temporal relationship (e.g., compartmentalisation of nuclear functions is particularly seen with replication), DNA moves through a fixed architec-

ture containing the molecular machines directing replication.

- **nucleus breeding**
  breeding scheme where populations in breeding cycle are divided into intensively managed nucleus with top-ranking genotypes and less intensively managed genetically less advanced main population.

- **null allele**
  an allele whose effect is either an absence of normal gene product at the molecular level or an absence of normal function at the phenotypic level.

- **null hypothesis**
  the statistical hypothesis that states that there are no differences between observed and expected data.

- **null hypothesis test**
  the standard hypothesis used in testing the statistical significance of the difference between the means of samples drawn from two populations, the null hypothesis states that there is no difference between the populations from which the samples are drawn, one then determines the probability that one will find a difference equal to or greater than the one actually observed. If this probability is 0.05 or less, the null hypothesis is rejected and the difference is said to be statistically significant.

- **null mutation**
  a mutation that eliminates all enzymatic activity, usually deletion mutations.

- **nulli(somic)-tetrasomic line**
  nulli-tetrasomic.

- **nulli-haploid**
  a cell or individual that possesses a haploid chromosome set plus a missing single chromosome.

- **nulliplex type**
  the condition in which a polyploid carries a recessive gene at a particular locus in all homologues, simplex denotes that the dominant gene is represented one, duplex two, triplex three, quadruplex four times, etc.

- **nullisome**
  a plant lacking both members of one specific pair of chromosomes.

- **nullisomic**
  a diploid cell or individual missing both copies of the same chromosome. A cell or individual with one chromosomal type missing, with a chromosome number such as n-1 or 2n-2.

- **nullisomic analysis**
  in nullisomic analysis, observations are made for phenotypic or other differences between the nullisomic for each chromosome and the disomic condition within the same variety. The method is applied for localisation of genes within a given genome. The method can only be used in polyploids, while diploids commonly do not tolerate the loss of both homologous chromosomess.

- **nulli-tetrasomic**
  a cell or individual, usually an allopolyploid, that possesses one lacking pair of chromosomes, which is partially compensated by a tetrasomic (four-fold dosage) of another. Usually homoeologous chromosome, a whole series of nulli-tetrasomics was produce in hexaploid wheat and successfully used in numerous genetic and molecular studies.

- **numeric aberration**
  the variation of the number of genomes or chromosomes, for example, ploidy variation, aneuploids (nullisomics, monosomics, trisomics, tetrasomics, etc.), substitutions or additions.

- **numeric constancy of chromosomes**
  the constant inheritance of the same number of chromosomes from generation to generation, which is facilitated by the mitotic and meiotic mechanisms.

- **numerical taxonomy**
  the classification of related organisms using a multitude of characteristics, each one of which is given equal weight, the degree of similarity between them is calculated using a digital computer, which treats the data collected for all characters and determines the similarities taking all possible pairs tax.

- **nurse crop**
  companion crop.

- **nursery**
  a place where young trees or other plants are raised, either for propagation or for testing and observations.

- **nurseryman's tape**
  grafting tape.

- **nut**
  a dry, indehiscent, woody fruit.

- **nutation**
  the turning of a plant or plant organ toward light.

■ **nutlet**
a little nut (e.g., in strawberry) nut.

■ **nutrient**
a nutritive substance or ingredient, such as major and minor mineral elements, necessary for plant growth and development as well as the organic addenda such as sugars, vitamins, amino acids and others employed in plant tissue culture media.

■ **nutrient-enhanced varieties (crops)**
plants that have been modified to possess novel traits that make those plants more economically valuable for nutritional uses (e.g., higher than normal protein content in feedgrains, high-zinc content or high-glutenin in wheat, high-amylose, high-lysine, high-methionine high-oil in maise, high-phytase in maize and soybeans, high-oleic oil, high-stearate, high-sucrose in soybeans.)

■ **nutritional mutant**
nutrient-enhanced varieties.

■ **nutritive substance**
nutrient.

■ **nutritive value index**
a measure for daily digestible amount of forage per unit of metabolic body size relative to a standard forage. It is used as a selection criterion in forage crop breeding.

■ **nyctanthous**
flowering by night.

■ **obligate apomixis**
seed apomixis with maternal offspring to 100 percent.

■ **ocdna**
an open circular DNA molecule that has at least one nick, it cannot be supercoiled and has the same density as linear DNA in $CsCl_2$/EB density gradients.

■ **ochre codon**
the codon UAA, a nonsense codon.

■ **octad**
an ascus containing eight ascospores, produced in species in which the tetrad normally undergoes a postmeiotic mitotic division.

■ **octopine**
a rare derivate of an amino acid, it is produced in some crown galls of plants, the controlling genes are part of the t-DNA of Ti plasmids.

■ **octoploid**
having eight chromosome sets of identical or different complements.

- **octovalent**
  chromosome configuration consisting of eight chromosomes.
- **ocular**
  eyepiece.
- **off-grade**
  postharvest removal of pathogen-infested or damaged fruit, seeds or plants by screening procedures, the culled or off-graded material can later be individually analysed or discarded.
- **offset**
  a young plant produced by the parent, usually as its base (offshoot) or a small bulb at the base of a mother bulb.
- **offspring**
  progeny.
- **off-type**
  an individual differing from the population norm in morphological or other traits, the term also includes escapes and contaminants (e.g., seeds that do not conform to the characteristics of a variety, uncontrolled self-pollination during production of hybrid seed, segregates from plants, etc.)
- **oil crops**
  plants that are grown for oil or oil-like products, main oil crops are: castor, peanut, rapeseed, safflower, sesame, soybean, sunflower, crambe, niger, jojoba and poppy.
- **oil legumes**
  oil crops.
- **oil poppy**
  oil crops.
- **oilseed**
  oil crops.
- **okazaki fragment**
  segment of newly replicated DNA produced during discontinuous DNA replication.
- **olein**
  a colourless to yellowish, oily, water-insoluble liquid, $C_{57}H_{104}O_6$, the triglyceride of oleic acid, present in many vegetable oils.
- **oligogene**
  a gene that produces a pronounced phenotypic effect on characters that show normal inheritance.
- **oligogenic**
  inheritance due to a small number of genes with discernible effects.
- **oligogenic resistance**
  resistance controlled by one or a few genes.
- **oligomer**
  a protein composed of two or a few identical polypeptide subunits.

■ **oligonucleotide (oligos)**
a small piece of ssDNA or ssRNA, oligos are synthesised by chemically linking together a number of specific nucleotides, they are used as synthetic genes and DNA probes or in site-directed mutagenesis.

■ **oligopeptide**
a small protein composed of 5-20 amino acids.

■ **oligospermous**
showing only few seeds.

■ **omnipotency**
totipotency.

■ **omnivorous**
of parasites or attacking a number of different hosts.

■ **once blooming**
it refers to plants or varieties (e.g., in roses) that bloom once a year.

■ **one-gene-one-enzyme (polypeptide) hypothesis**
the hypothesis that a large class of structural genes exists in which each gene encodes a single polypeptide that may function either independently or as a subunit of a more complex protein. Originally it was thought that each gene encoded the whole of a single enzyme, but it has since been found that some enzymes and other proteins derive from more than one polypeptide and hence from more than one gene.

■ **one-point crossover**
in genetic algorithms, a breeding technique using one randomly chosen point, interchanging the portions of the two breeding individuals to the right of that point.

■ **ontogenesis**
the course of growth and development of an individual from zygote formation to maturity.

■ **ontogeny**
ontogenesis.

■ **oocyte**
the gamete in females. See **ovum**.

■ **oogonia**
cells in females that produce primary oocytes by mitosis.

■ **oogonium**
a primordial germ cell that give rise, by mitosis, to oocytes, from which the ovum and polar bodies develop by meiosis.

■ **opal**
a UAG stop codon nonsens mutation.

■ **opal codon**
the codon UGA, a nonsens codon.

- **opaque**
  partially pervious to light.
- **opaque (–2) maize (mutant)**
  a mutant form that produces proteins rich in lysine and higher in content of calcium, magnesium, iron, zinc and manganese.
- **open pollination**
  natural, cross or random pollination, a free gene flow.
- **open storage**
  storage with free access to normal atmospheric conditions.
- **open-pollinated crossing group**
  crossing group(s).
- **open-pollinated variety**
  a variety multiplied through random fertilisation, as opposed to a *hybrid* variety.
- **operator**
  a DNA region at one end of an operon that acts as the binding site for repressor protein. A DNA sequence that is recognised by a repressor protein or repressor-corepressor complex. When the operator is complexed with the repressor, transcription is prevented.
- **operon**
  a set of adjacent structural genes in bacteria whose mRNA is synthesised in one piece, together with the adjacent regulatory signals that affect transcription of the structural genes. A sequence of adjacent bacterial genes all under the transcriptional control of the same operator.
- **opposite**
  applied to the leaf arrangement in which leaves arise in pairs, one pair at each node.
- **optic chiasma**
  it refers to a visible chiasma on meiotic chromosomes through the microscope.
- **optimal sampling strategy**
  a sampling strategy that ensures that the genetic diversity of a species is represented in the samples.
- **opuntia**
  within the subtribe *Opuntioideae* there are several species used as crop and horticultural plants, edible fruits and fleshy parts of the plant, special use for production of the stain 'carmine red' by the ecto-parasite Cochenille *(Dactylopius coccus)*.

- **orchard**
an area of land devoted to the cultivation of (fruit or nut) trees.

- **ordinal scale**
a scale for scoring quantitative data using a series of predefined intervals arranged in a logical sequence (e.g., a typical ordinal scale may involve responses of 'very good,' 'good,' 'satisfactory,' 'poor,' or 'very poor').

- **organ asymmetry**
in many plants the left and right halves of their organs have distinct shapes (e.g., the leaves of *Begonia, Tilia* (lime tree), *Ulmus* (elm) and petals of *Anhirrhinum* (snapdragon) or *Pisum* (pea) flowers, they can occur in two mirror-image forms, left handed and right-handed. In many cases these two forms occur in equal numbers on the plant, either being located opposite each other or alternating along the stem, asymmetry of each organ traces back to a meristem with a single plane of symmetry (bilateral symmetry), such that mirror-image organs arise from opposite halves of the meristem.

- **organ culture**
the growth in aseptic culture of plant organs, such as roots or shoots, beginning with organ primordia or segments and maintaining the characteristics of the organ.

- **organelle**
1. a subcellular structure having a specialised function for example the mitochondrion, the chloroplast or the spindle apparatus.
2. within a cell, a persistent structure that has a specialised function, mostly separated from the rest of the cell by selective membranes.

- **organic soil**
a soil that is composed predominantly of organic matter, it usually refers to peat.

- **organogenesis**
the initiation and growth of an organ from cells or tissue.

- **organogenetic**
it refers to cells or tissue able to form organs.

- **organogenic**
organogenetic.

- **origin of replication**
the point of specific sequence at which DNA replication is initiated.

- **ornamental plant**
a plant that is grown for visual display.

- **ornithogamy**
  ornithophily.

- **ornithophily**
  pollination by birds.

- **orphan gene**
  a gene identified by sequencing, its function is unknown.

- **ortet**
  the original plant from which a clone is started through rooted cuttings, grafting, tissue culture or other means of vegetative propagation (e.g., the original plus tree used to start a grafted clone for inclusion in a seed orchard.)

- **orthodox seed**
  seed that can be dried and stored for long periods at reduced temperatures and under low humidity.

- **orthologous genes**
  homologous genes that have become differentiated in different species derived from a common ancestral species, as opposed paralogous genes.

- **orthoploid**
  euploid.

- **oryzalin**
  an agent that is efficient for chromosome doubling of haploid apple shoots.

- **oryzinin**
  glutenin.

- **osmolarity**
  the total molar concentration of the solutes affecting the osmotic potential of a solution or nutrient medium.

- **osmolyte**
  osmolytes are osmolytic active, neutral organic compounds, such as sugars (polyols), certain amino acids and quaternary ammonium compounds. Proline is the most widely distributed compatible osmolyte. There is a strong correlation between increased cellular proline levels and the capacity to survive both water deficiency and the effects of high environmental salinity.

- **osmosis**
  the net movement of water or of another solvent from a region of low solute concentration to one of higher concentration through a semipermeable membrane.

- **other crop seed**
  one of the four components of a purity test, the total percentage (by weight) of seed of all crop species, each comprising less than 5 percent of the seed lot.

- **o-type**
  a maintainer plant in sugarbeet breeding, it carries the same ste-

rility genes as the male sterile plants but having the normal cytoplasm—(N)xxzz. This genotype exists at low frequencies (3-5 percent) in most sugarbeet populations. It can be identified only by test-crossing prospective O-types with CMS plants. If all the offspring from a test cross are male sterile, the test-crossed pollinator plant is of the O-type genotype. By repeated selfing of an identified O-type and simultaneous repeated backcrossing to a CMS line, inbred O-type lines and their equivalent inbred CMS lines can be developed.

- **outbreeding**
  the crossing of plants that are not closely related genetically, in contrast to inbreeding, in which the individuals are closely related.

- **outclassed**
  it refers to a crop variety that is taken away from registration.

- **outcrossing**
  cross pollination between plants of different genotypes. In biotechnology, the transfer of a given gene or genes (e.g., one synthesised by humans and inserted into a plant via genetic engineering) from a domesticated organism (e.g., crop plant) to wild type (relative of crop.)

- **outgrades (in potato)**
  outgrades are tubers considered unmarked because of size, disease, greening, second growth or slug or mechanical damage.

- **outlier**
  an individual that occurs naturally some distance away from the principal area in which its population is found. They are anomalous values in the data and can be due to recording errors, which may be correct or they may be due to the sample not being entirely from the same population.

- **outplant**
  a seedling, transplant or cutting ready to be established on a certain site.

- **outside markers**
  loci on either side of another locus or specified region.

- **ovary**
  the part of the flower that develops into the grain in grasses, the ovary has one or more ovules, each containing an embryo sac.

- **ovary wall**
  pericarp.

- **ovate**
  egg-shaped, having an outline like that of an egg, with the broader end basal ovoid.

■ **overall resistance**
resistance to disease expressed at all plant growth stages.

■ **overdominance**
a relationship in which the phenotypic expression of the heterozygote is greater than that of either homozygote.

■ **overhang**
3' and 5' ssDNA overhangs of dsDNA, overhangs may also be called extensions or sticky ends.

■ **overlapping code**
overlapping DNA (segments).

■ **overlapping DNA (segments)**
a special type of gene organisation, one DNA sequence may code for different proteins. It is performed by two open reading frames, which act subsequently.

■ **overseeding**
seeding into an existing crop stand or turf.

■ **overstored seeds**
method of overstored seeds.

■ **oversummering**
the survival through the summer and/or to keep alive through summer.

■ **overwintering**
the survival through the winter and/or to keep alive through winter.

■ **ovoid**
egg-shaped ovate.

■ **ovule**
a structure in angiosperms and gymnosperms that, after fertilisation, develops into a seed.

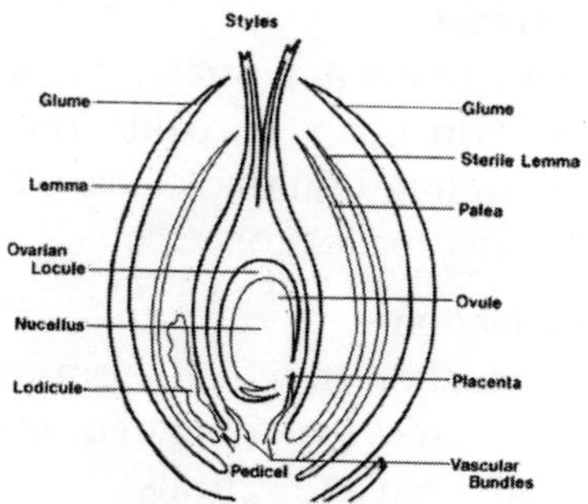

■ **ovule primordium**
meristematic tissue of the ovary wall from which the seeds of angiosperms originate.

■ **ovum**
egg. The one functional product of each meiosis in female animals.

■ **ovum**
egg.

■ **oxalate oxidase**
an enzyme detected in, for example, barley seedling roots soon after germination and in the leaves of mature plants and in response to powdery mildew infection, the enzyme contains manganese, the enzyme shows

almost identical structure to the wheat protein germin.

- **oxalic acid**
  a white, crystalline, water-soluble, poisonous acid, $H_2C_2O_{4\cdot2}H_2O$, used for bleaching and as a laboratory reagent.

- **oxidase**
  an enzyme that catalyses reactions involving the oxidation of a substrate using molecular oxygen as an electron acceptor.

- **oxidation**
  a reaction in which atoms or molecules gain oxygen or lose hydrogen or electrons.

- **oxidative burst**
  an early defence response against pathogens where 10^{-14} mol H_2O_2 per cell per second is generated, the early oxidative burst (phase I) is common for compatible and incompatible interactions, the later, but sustained (phase II) is associated with hypersensitive response/reaction.

- **p (peptidyl) site**
  the site on the ribosome occupied by the peptidyl-tRNA just before peptide bond formation.

- **p element**
  a Drosophila transposable element that has been used as a tool for insertion mutagenesis and for germline transformation.

- **P VALUE**
  Probability Value, a decimal fraction showing, e.g., the number of times an event will occur in a given number of trials.

- **p1**
  parental generation.

- **pachynema (pachytene stage)**
  the stage of prophase 1 of meiosis in which chromatids are first distinctly visible.

- **panmictic**
  referring to unstructured (random-mating) populations.

- **paracentric inversion**
  an inversion not involving the centromere. A chromosomal inversion that does not include the centromere.

- **paralogous genes**
  two genes or clusters of genes at different chromosomal locations in the same organism that have structural similarities indicating that they derived from a common ancestral gene.

- **parameters**
  measurements of attributes of a population.

- **parapatric speciation**
  speciation in which the evolution of reproductive isolating

mechanisms occurs when a population enters a new niche or habitat within the range of the parent species.

- **Parental Ditype (PD)**
a tetrad type containing two different genotypes, both of which are parental. A spore arrangement in Ascomycetes that contains only the two non-recombinant-type ascospores.

- **parthenogenesis**
the production of offspring by a female with no genetic contribution from a male. The development of an individual from an unfertilised egg that did not arise by meiotic chromosome reduction.

- **partial digest**
a restriction digest that has not been allowed to go to completion and thus contains pieces of DNA with some restriction endonuclease sites that have not yet been cleaved.

- **participatory plant breeding**
broadly defined as approaches that involve close collaboration between researchers and farmers and potentially other stakeholders, to bring about plant genetic improvements within a species. PPB covers the whole research and development cycle of activities associated with plant genetic improvement identifying breeding objectives, generating genetic variability or diversity, selecting within variable populations to develop experimental materials, evaluating these materials (this is known as *participatory variety selection* or PVS), release of materials, diffusion, seed production and distribution. It also could include assessing existing policy or legislative measures or both and designing new ones where needed. Farmers and breeders and other stakeholders — such as traders, processors and consumers — can take on different roles at various points in the cycle but they join forces to bring about change.

- **participatory variety selection**
the selection of fixed lines (including landraces) by farmers in their target environments using their own selection criteria. Consists of four methodological steps:
1. Situation analysis and identification of farmers' varietal needs.
2. Search for suitable genetic materials.

3. Farmers' experimentation with new crop varieties in their own fields and with their own crop-management practices.
4. Wider dissemination of farmer-preferred crop varieties.

- **pattern formation**
the developmental processes by which the complex shape and structure of higher organisms occurs.

- **PCR**
Polymerase Chain Reaction, a technique for copying the complementary strands of a target DNA molecule simultaneously for a series of cycles until the desired amount is obtained.

- **pedigree**
a family tree drawn with standard genetic symbols, showing inheritance patterns for specific phenotypic characters. A representation of the ancestry of an individual or family, a family tree.

- **penetrance**
the proportion of individuals with a specific genotype who manifest that genotype at the phenotype level.

- **peptide bond**
a bond joining two amino acids.

- **peptidyl transferase**
the enzymatic centre in the ribosome responsible for peptide bond formation during translation.

- **pericentric inversion**
an inversion that involves the centromere. A chromosomal inversion that includes the centromere.

- **permissive condition**
environmental condition under which a conditional mutation shows the wild-type phenotype.

- **permissive temperature**
a temperature at which temperature-sensitive mutants are normal.

- **persistence**
the act or fact of persisting.

- **persistent**
persistence.

- **persistent modification**
non-heri morphological or physiological changes over a more or less long period of generation cycles induced by varying abiotic or biotic influences.

- **pest**
any form of plant or animal life or any pathogenic agent, injurious or potentially injurious of plants or plant products.

■ **pest hypothesis**
the short life-time of a protein is signalled by a region rich in the amino acids proline (P), glutamic acid (E), serine (S) or threonine (T).

■ **pest resistance**
resistance to any form of plant or animal life or any pathogenic agent.

■ **pesticide**
a chemical preparation for destroying plant, fungal or animal pests.

■ **petal**
one of the inner floral leaves, usually brightly coloured and borne in a tight spiral or whorled carolla.

■ **petiole**
the stalk by which a leaf is attached.

■ **petite**
a yeast mutation producing small colonies and altered mitochondrial functions. In cytoplasmic (neutral and suppressive) petites the mutation is a deletion in mitochondrial DNA, while in segregational petites the mutation occurs in nuclear DNA.

■ **petite mutation**
a mutation of yeast that produces small (petite) anaerobic-like colonies.

■ **ph locus**
a gene that controls homoeologous chromosome pairing in wheat and, similarly, in other allopolyploid plants, the *Ph* gene restricts pairing between homologous chromosomes in a polyploid.

■ **pH value**
the negative logarithm of the hydrogen-ion activity, expressed in terms of the pH scale from 0-14.

■ **phage**
the abbreviation for bacteriophage (i.e., a virus that infects bacteria).

■ **phasmid**
a cloning vector that has the possibilities to replicate as a plasmid or as a phage, the two modes of replication are usually functional in different bacterial species.

■ **PHE**
phenylalanine (an amino acid).

■ **phellogen**
a layer of plant tissue outside of the true cambium, giving rise to cork tissue.

■ **phene**
the phenotype of the plant, which is a product of the gene (gene > DNA > transcription >

RNA > processing > translation > protein) and the interaction with the environment phenotype.

- **phenocopy**
  a phenotype that is not genetically controlled but looks like a genetically controlled phenotype. An environmentally induced phenotype that resembles the phenotype produced by a mutation.

- **phenogenetics**
  a branch of genetics that studies the interaction of the genotype and its manifestation.

- **phenol**
  a chemical used to remove proteins from DNA preparations.

- **phenol test (reaction)**
  the colour produced in the grain of, for example, wheat and barley, by treatment with a one percent solution of phenol in water.

- **phenology**
  the study of the impact of climate on the seasonal occurrence of flora.

- **phenotype**
  the observable manifestation of a specific genotype (i.e., those properties on an organism, produced by the genotype in conjunction with the environment).

- **phenotypic expression**
  the manifestation of a particular gene resulting in a particular phenotype.

- **phenotypic plasticity**
  the capability of a genotype to assume different phenotypes.

- **phenotypic segregation**
  the phenotypic differentiation patterns of cells or individuals in segregating populations, as opposed to genetic segregation.

- **phenotypic selection**
  development of a variety based on its physical appearance without regard to its genetic constitution.

- **phenotypic sex determination**
  sex determination by non-genetic means.

- **phenotypic variance**
  variance of phenotype due to genotypic and environmental factors combined.

- **Phenotypic Variance (VP)**
  the total variance observed in a character, it includes experimental error, genotype × environment interaction and the genotypic variance.

- **phenylalanine (phe)**
  an aromatic, nonpolar amino acid.

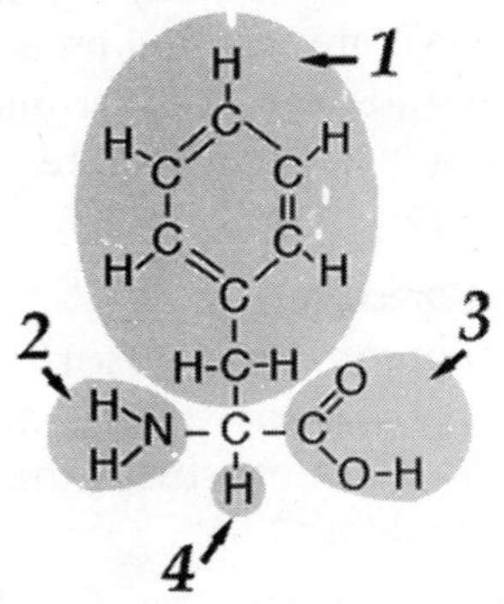

- **pheromone**
  a chemical signal, analogous to a hormone, that passes information between individuals.

- **phloem**
  a tissue comprising various types of cells that transports dissolved organic and inorganic materials over long distances within vascular plants.

- **phosphatase**
  an enzyme that catalyses reaction involving the hydrolysis of esters of phosphoric acid.

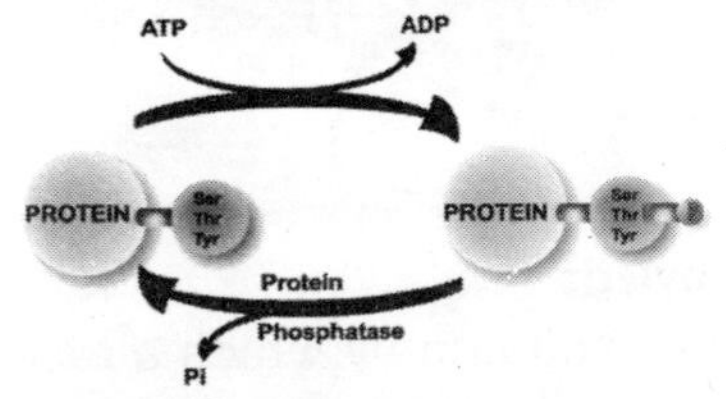

- **phosphodiester bond**
  a bond between a two sugar groups and a phosphate group, such bonds form the sugar-phosphate-sugar backbone of DNA and RNA. A diester bond (between phosphoric acid and two sugar molecules) linking two nucleotides together to form the nucleotide polymers DNA and RNA.

- **phosphorus (p)**
  an element that is required by plants in the oxidised form, it is utilised in reactions in which energy is transferred, often involving ATP mitochondrion.

- **phosphorylation**
  the addition of a phosphate group to a compound, involving the formation of an ester bond between the reactants.

- **photobleaching**
  photochemical reaction of fluorophores, light and oxygen that causes the intensity of the fluorescence emission to decrease with time.

- **photoinhibition**
  the slowing or stopping of a plant process by light (e.g., the germination of some seeds.)

- **photo-insensitive plants**
  daylength insensitivity.

- **photometry**
  the measurement of the intensity of light or of relative illuminating power.

▪ **photomorphogenesis**

changes in plant growth due to light. There is a main plant regulator protein Cop1 that suppresses genes controlling the photomorphogenesis, for example, when a seedling gets exposed to light the Cop1 protein is reduced in the nucleus and photosynthesis is initiated—the seedling becomes green, cryptochromes may interact with photoreceptor-proteins, which can recognise blue light, thus the interaction can 'switch-off' the Cop1 protein.

▪ **photon**

a packet of light energy.

▪ **photoperiod**

the relative length of the periods of light and darkness associated with day and night, in many species, floral induction occurs in response to daylength, species have been categorised according to their daylength requirements as short-day, long-day, intermediate-day or day-neutral.

▪ **photoperiodism**

the response of a plant to periodic, often rhythmic, changes in either the intensity of light or to the relative length of day.

▪ **photoreactivation**

the process whereby dimerised pyrimidines (usually thymines) in DNA are restored by an enzyme (deoxyribodipyrimidine photolyase) that requires light energy.

▪ **photoreceptor**

a pigment that absorbs the light used in various metabolic plant processes that require light.

▪ **photosynthesis**

the series of metabolic reactions that occur in certain autotrophs, whereby organic compounds are synthesised by the reduction of carbon dioxide using energy absorbed by chlorophyll from light.

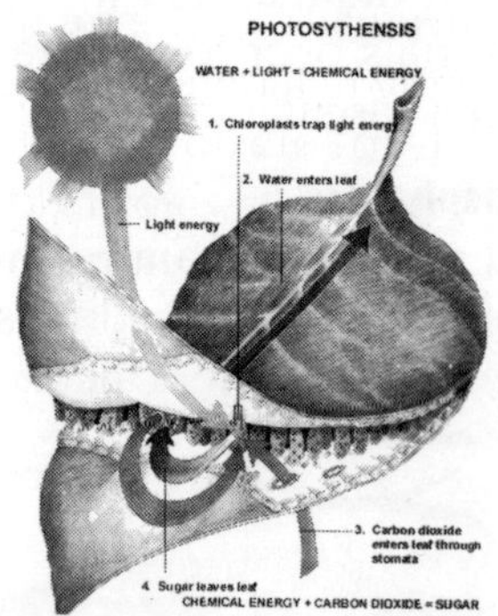

▪ **phyletic**

the evolution by which a race or line is progressively transformed from its ancestral form without branching or separating into related parts.

- **phyletic gradualism**
  the process of gradual evolutionary change over time.
- **phyletic series**
  phyletic.
- **phyllode**
  an expanded petiole resembling and having the function of a leaf, but without a true blade.
- **phyllody**
  the condition in which parts of a flower are replaced by leaf-like structures, often it is a symptom of certain diseases.
- **phyllosphere**
  the surface of a living leaf.
- **phyllotaxis**
  the active arrangement of leaves on a stem or axis.
- **phylogenesis**
  phylogeny.
- **phylogenetic**
  of, pertaining to or based on phylogeny.
- **phylogenetic tree**
  a diagram showing evolutionary lineages of organisms.
- **phylogeny**
  evolutionary relationships within and between taxonomic levels, particularly the patterns of lines of descent, often branching, from one organism to another.
- **phylum**
  an important group of organisms.
- **physical map**
  a map of the locations of identifiable landmarks on DNA (e.g., *restriction enzyme cutting sites, genes*), regardless of inheritance. Distance is measured in *base pairs*.
- **physiologic specialisation**
  the existence of a number of races or forms of one species of pathogen based on their pathogenicity to different cultivars of a host.
- **physiological maturity**
  the maturity of a seed when it reaches its maximum dry weight, this usually occurs prior to the normal harvest date.
- **physiological race**
  pathogens of the same species with similar or identical morphology but differing pathogenic capabilities.
- **physiologist**
  a specialist dealing with the functions and activities of living organisms and their parts.
- **phytase**
  an enzyme that catalyses the breakdown of phytin, the source of inorganic phosphorus in seed metabolism.

- **phytoalexin(e)**
an antifungal substance that is produced by a plant in response to damage or infection.

- **phytochrome**
a photoreversible pigment that occurs in every major taxonomic group of plants. It exists in two interchangeable forms with respect to absorption, a red and a far-red form, Pr phytochrome is receptive to orange-red light (600-680 nm) and inhibits flowering, Pf-r phytochrome is receptive to far-red light (700-760 nm) and induced flowering.

- **phytogenetics**
synonymous to plant genetics, dealing with inheritance in plants.

- **phytohormone**
plant hormone.

- **phytoncide**
See **herbicide.**

- **phytopathology**
the study of plant diseases.

- **phytosiderophore(s)**
nonproteinogenic amino acids developed by plants under conditions of mineral deficiency (especially under iron and zinc deficiency), the production and exudation of phytosiderophores is controlled by several genes, there are crop plants, such as rye, showing a high level of phytosiderophore production and/or exudation toward the rhizosphere.

- **phytotoxic**
being poisonous to plants.

- **phytotoxin**
pathotoxin.

- **phytotron**
a group of rooms or a room for growing plants under controlled and reproducible environmental conditions.

- **pick-up reel**
a special device on some harvesters for taking up lodging straw.

- **pigment**
an organic compound that produces colour in the tissue of the plant.

- **pileorhiza**
root cap.

- **piliferous layer**
root hair.

- **pilosum**
used for describing a hairy ventral furrow, for example, barley grain.

- **pilus**
a conjugation tube, a hollow hair-like appendage of a donor *Escherichia coli* cell that acts as

a bridge for transmission of donor DNA to the recipient cell during conjugation.

- **pin flower**
flowers with long styles and short stames.

- **pincers**
a gripping tool consisting of two pivoted limbs forming a pair of jaws and a pair of handles.

- **pinching (pinching back, pinching out)**
disbud.

- **pinna**
one of the leaflets of a pinnate leaf.

- **pinnate**
feathered.

- **pinnate leaf**
pinnate.

- **pinninervate**
pinnate-veined, feather-veined.

- **pistil**
the gynoecium of a syncarpous flower, a pistil includes an ovary, style and stigma, the stigma is the receptor of the pollens.

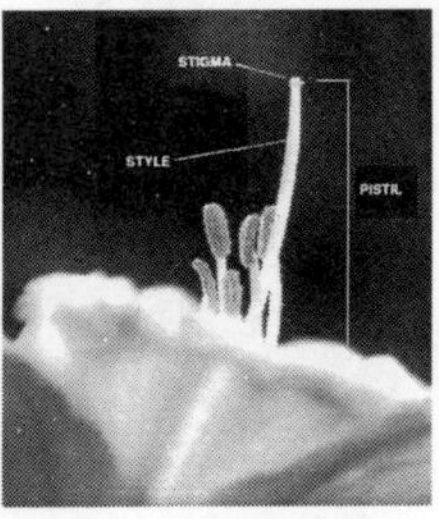

- **pistillate flower**
designating a flower having one or more pistils and no stamens.

- **pistillody**
the conversion of any organ of a flower into carpels (e.g., stamens into pistils or pistillike structures) pistillate flower.

- **pistillum**
pistil.

- **pit**
in botany, a term used for the widening in the center of the ventral furrow in some wheat and barley grains.

- **pith**
a tissue that occupies the central part of a stem (composed of parenchyma cells).

- **plant (geno) type**
the entire genetic constitution of a plant variety. An off-type is a plant differing from the variety in morphological or other traits.

- **plant breeding**
the application of genetic principles and practices to development of individuals or cultivars more suited to the needs of humans, it uses knowledge from agronomy, botany, genetics, cytogenetics, molecular genetics, physiology, pathology, entomology, biochemistry, statistics, etc.

- **plant density**
the rate at which seed or vegetative propagules are placed in a field or experimental planting.

- **plant hair**
trichome.

- **plant hormone**
a compound that is synthesised by a plant, but is not a nutrient, coenzyme or detoxification product and which regulates growth, differentiation or other specific physiological processes.

- **plant label**
plastic, wood or other stakes for gardens or experiments to indicate what seeds or material are planted where until they appear, which varieties are included, what sort of evaluation is carried out, etc., in plant conservation, paper forms to include in drying plant samples, with formal printed forms as permanent labels on herbarium specimens. The minimum information includes the name of the collector, the location collected, the date collected and the correct identification of the specimen.

- **plant passport**
an official seed label used for forthcoming marketing, it shows the crop, crop class, e.g., European community grade, inspections, etc. passport data.

- **Plant Pathogenesis-Related Proteins (PPRP)**
groups of proteins with different chemical properties produced in a cell within minutes or hours following inoculation, but all being more or less toxic to pathogens.

- **plant pathology**
phytopathology biological control.

- **plant protection**
phytopathology biological control.

- **plant species**
a group of organisms capable of interbreeding freely with each other but not with members of other species. In taxonomic classification, a subdivision of a genus, a group of closely related individuals descended from the same stock.

- **plant variety**
in classical botany, a variety is a subdivision of a species. An agricultural variety is a group of similar plants that by structural features and performance can be identified from other varieties within the same species.

- **plantation**
  a closely set stand of trees or special crops that has been planted by humans.

- **plantibodies**
  antibodies produced in transg-enic plants expressing the antibody-producing gene(s) of an animal, e.g., mouse, that had been previously injected with a pathogen (usually a virus) and that infects the plants.

- **planting cord**
  a string of different manufacturing and length used for marking experimental plots, paths. margins between landmarks or applied to mark planting rows.

- **plantlet**
  a stage of in vitro culture, the stage after torpedo stage and usually one of the last before a whole plant is generated.

- **plantling**
  See **plantlet**.

- **plaque**
  a clear area on a bacterial lawn, left by lysis of the bacteria through progressive infections by a phage and its descendants. Clear area on a bacterial lawn caused by cell lysis due to viral attack.

- **plaque lift**
  impression of bacteriophage particles blotted to a filter membrane, the technique is used to screen both with radioactively-labelled DNA or RNA probes or with serological screening of eukaryotic expression libraries.

- **plasmagene**
  an extranuclear hereditary determinant showing no Mendelian inheritance.

- **plasmalemma**
  cell membrane.

- **plasmatic**
  all functions, processes or properties of the cytoplasm cytoplasm.

- **plasmic**
  See **plasmatic**.

- **plasmid**
  autonomously replicating, extrachromosomal circular DNA molecules, distinct from the normal bacterial genome and nonessential for cell survival under nonselective conditions. Some plasmids are capable of integrating into the host genome. A number of artificially constructed plasmids are used as cloning vectors.

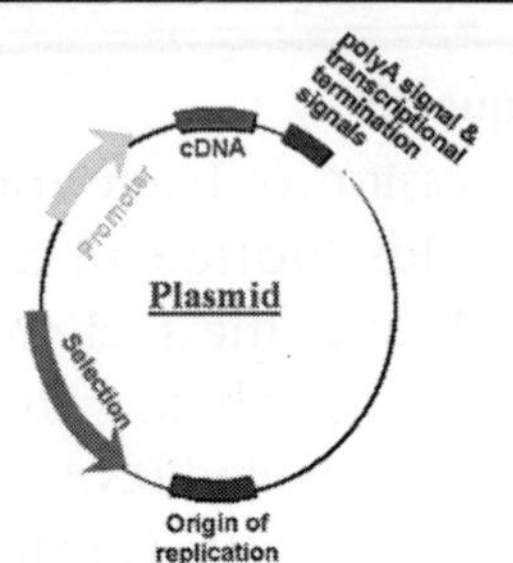

- **plasmid vector**
a plasmid or plasmidlike structure used as a carrier for alien DNA segments or genes.

- **plasmodesm(a)**
cytoplasmic bridges, lined with a plasma membrane that connect adjacent cells. They provide major pathways of communication and transport between cells.

- **plasmodium**
in cellular slime moulds, a vegetative structure consisting of a noncellular, mobile mass of naked protoplasm containing many nuclei.

- **plasmogamy**
the fusion of the cytoplasm of two or more cells after karyogamy during the process of fertilisation.

- **plasmolysis**
the result of placing plant cells in a hypertonic solution so that water is drawn out of the cell, the cytoplasm shrinks and the cell membrane is pulled away from the cell wall.

- **plasmon**
all the extrachromosomal hereditary determinants plasmotype.

- **plasmon mutation**
a mutation that genetically changes the cytoplasm and/or its hereditary determinants.

- **plasmotype**
the sum of the extrachromosomal hereditary determinants.

- **plastid**
1. a chloroplast prior to the development of chlorophyll.
2. one of a group of double-membrane-bound plant-cell organelles that vary in their structure and function (e.g., chloroplasts, leucoplasts, amyloplasts).

- **plastid DNA**
organelle DNA that is present in a plastid.

- **plastid inheritance**
non-Mendelian inheritance that is caused by hereditary factors present in the plastids.

- **plastidome**
See **plastome**.

- **plastidome mutation**
mutations that genetically change the DNA of the plastids of a plant cell.

- **plastogene**
  the hereditary determinants located in the plastid.

- **plastom mutation**
  plastidome mutation.

- **plastome**
  a term usually used for the plastids of a cell or for the genetic information of the plastid DNA.

- **plate**
  to place on or in special media in a culture dish.

- **plating efficiency**
  an estimate of the percentage of viable cell colonies developing on an agar plate relative to the total number of cells spread onto the plate. The plating efficiency is a function of the tissue, medium composition, plating density and the phase of the stock culture.

- **pleiotropic**
  an allele or gene that affects several traits at the same time.

- **pleiotropic mutation**
  a mutation that has effects on several different characters.

- **pleiotropy**
  the phenomenon of a single gene being responsible for a number of different phenotypic effects that are apparently unrelated.

- **plesiomorphy**
  ancestral state of a particular character.

- **ploidy**
  the number of complete chromosome sets in the cell nucleus (e.g., diploid, tetraploid, etc.).

- **plot**
  in field experiments, more or less large pieces of land used for planting and evaluation, in forestry, a group of trees, all from the same entry (family, clone, provenance) planted together. A five-tree-plot row is the most common design for forest genetics experiments.

- **plot size**
  in field experiments, the size, number and distribution of plots are essential elements. An efficient combination of plot size and plot number is required. Larger plots offer smaller between-plot variance, long measurement time per plot, shorter walking time between plots, less edge and less statistical error. Small plots require a higher number of plots in order to achieve the same level of precision of estimates, plot size and shape vary with crops and stage of testing and among characters under selection, plots are generally small at the initial stage of

testing and reach a maximum size during the second and third years of replicated trials, unbordered plots with few replications at one or few locations are used for traits with low heritability, row spacing and plant populations are chosen to be similar to commercial production of crop. Seedbed preparation, fertilisation, weed control practices, etc., are generally the same as those used for commercial production. The mechanisation and computerisation of most plant breeding programmes have greatly increased a breeder's ability to handle more plots, populations, etc.

- **plumula**

the undeveloped shoot consisting of unexpanded leaves and the growing point (i.e., the terminal bud of developed embryo).

- **plurannual**

perennial.

- **plus tree**

a tree phenotype judged (but not proved by testing) to be unusually superior in some qualities (e.g., growth rate relative to site, growth habit, high wood quality, resistance to disease and insect attack or to other adverse local factors.)

- **pneumatophore**

a specialised root in certain aquatic plants that performs respiratory functions.

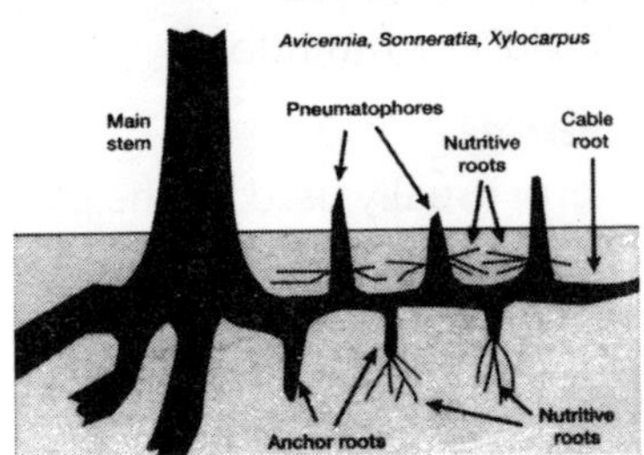

- **pod**

a fruit that dehisces down both sides into two separate valves that are most typically dry and somewhat woody, they are characteristic fruits of the Leguminosae legume.

- **pod drop**

losses due to the premature drop of pods,.

- **pod shattering**

seed losses due to the premature shattering of pods, for example, in oilseed rape it can be as great as 50 percent of the potential yield in some seasons. Average losses are around 10-15 percent, equivalent to 500 kg/ha or ten times the sowing rate.

- **point mutation**

a mutation that can be mapped to one specific site within a locus. A small mutation that consists of the replacement (transition or transversion), addition or deletion (frameshift) of one or a few bases.

- **poison**
toxicity.

- **poisson distribution**
a mathematical expression giving the probability of observing various numbers of a particular event in a sample when the mean probability of that event on any one trial is very small.

- **poky mutation**
a mutation of neurospora that produces a petite-like phenotype.

- **polar gene conversion**
a gradient of conversion frequency along the length of a gene.

- **polar granules**
cytoplasmic granules localised at the posterior end of a Drosophila oocyte and early embryo. These granules are associated with determinants of the germline and posterior structures.

- **polar mutation**
a mutation that affects the transcription or translation of part of the gene or operon downstream of the mutant site. For example, nonsense mutations, frameshift mutations and insertion sequence(IS)-induced mutations.

- **polar nuclei**
pole nucleus.

- **polarity**
meaning directionality and referring either to an effect seen in only one direction from a point of origin or to the fact that linear entities (such as a single strand of DNA or RNA or a protein) have ends (5' and 3', N and C) that differ from each other.

- **polarity gene**
a mitochondrial gene with alleles that are preferentially found in daughter mitochondria after a recombinational event between mitochondria.

- **polarisation**
restriction of the orientation of the vibration of electro-magnetic waves of light.

- **pole**
one of the two ends of the cell spindle toward which chromosomes move during mitotic and meiotic anaphase.

- **pole nucleus**
the two haploid nuclei present in the centre of the embryo sac after division of the megaspore. They may fuse to form a diploid definitive nucleus before fusing with the male gamete to form

the triploid primary endosperm nucleus prior to double fertilisation.

- **pollard**
  a tree cut back almost to the trunk in order to form a thick head of spreading branches, which are cut for basket-making and kindling (e.g., poplars and willows can be pollarded.)

- **pollen**
  collectively, the mass of microspores or pollen grains produced within the anthers of a flowering plant. It is a highly specialised tissue whose function is the production of two sperm cells and their subsequent delivery through the style and ovary to the embryo sac cells where the double fertilisation takes place. It is a highly reduced structure consisting of only three cells, the processes of pollen development, germination and fertilisation involve the specific expression of a large number of genes.

- **pollen analysis**
  a method to study pollen grains, particularly their size, shape and surface, since those characters are highly specific for species the method is used for taxonomic classifications.

- **pollen barrier**
  in seed production, the separation of the varieties, lines, etc. (mostly in allogamous crops) in order to prevent intercrossing. It is done by different means, such as strips of other plants (e.g., hemp or maize), which prevent the free flow of pollen or by isolation walls of tissue.

- **pollen contamination**
  refers often in connection with seed orchards to the phenomenon that some of the pollen fertilising orchard genotypes origins outside the seed orchard.

- **pollen culture**
  microspore culture.

- **pollen embryoids**
  embryoids that derive from anther culture.

- **pollen grain**
  a microspore in flowering plants that germinates to form the male gametophyte, a structure made up of the pollen grain plus a pollen tube pollen.

- **pollen grain mitosis**
  in microsporogenesis, a nuclear division that occurs in the pollen grain after the formation of tetrads, it results

in a smaller generative nucleus and a larger vegetative nucleus.

- **pollen mixing**
  cross-incompatibility in interspecific crosses is associated with proteins of the pistil that interact with proteins of the pollen to prevent normal pollen tube germination and growth. This unfavourable reaction may be avoided in certain combinations by mixing pollen from a compatible species with pollen from an incompatible parent.

- **Pollen Mother Cell (PMC)**
  the microsporocyte, which undergoes two meiotic divisions to produce four microspores, each microspore becomes a pollen grain.

- **pollen parent**
  the parent that furnishes the pollen and that fertilises the ovules of the other parent in the production of seed.

- **pollen sac**
  a sac within the anther of a stamen within which microspores are produced.

- **pollen sterility**
  pollen that is not able to fertilise an egg cell.

- **pollen transfer**
  refers to the kind of pollen transfer from the male to female organs (e.g., mediated by wind, insects, by hand, etc.)

- **pollen tube**
  the tube formed from a germinating pollen grain and down which the two male gametes pass to the ovum.

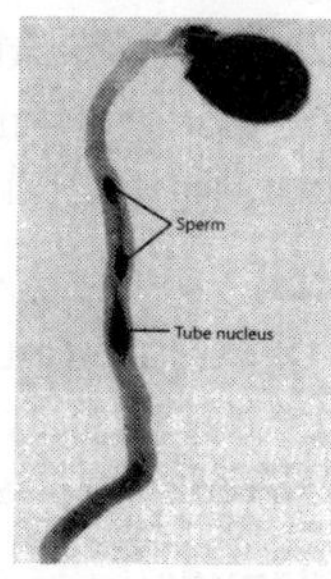

- **pollen tube competition**
  certation.

- **pollen-shedding**
  the status or process when the pollen grains are released from the anthers.

- **pollinarium**
  a functional unit in orchid pollination that consists of two or more pollinia, stalk or stipe and a viscidium.

- **pollinate**
  to transfer pollen from the anther to the receptive surface of the stigma of the same or another flower in angiosperms and from male to female in

gymnosperms. This process usually requires a vector in outbreeding plants.

- **pollination**
the transfer of pollen grains from the anther to the stigma of a flowering plant, pollination can also be done in artificial culture (i. e., in vitro pollination.)

- **pollinator**
that parental individual, line, variety or species, which is used as a donor of pollen in a cross.

- **poly(a) tail**
a string of adenine nucleotides added to the 3' end of eukaryotic mRNA after transcription.

- **polyacrylamide**
a material used to make gels for separation of mixtures of macromolecules by electrophoresis.

- **polyacrylamide gel**
a gel used to separate biological molecules (proteins), it is prepared by mixing a monomer (acrylamide) with a cross-linking agent (*N,N'*-methylene-bisacrylamide) in the presence of a polymerising agent. It leads to the formation of an insoluble three-dimensional network of monomer chains, which become hydrated in water.

- **polyacrylamide gel electrophoresis (page)**
a method for separation of proteins and amino acids on the basis of their molecule size, the molecules move through the gel under the influence of an electric field.

- **polyadenylation**
post-transcriptional addition of 50-200 adenine residues to the 3' end of eukaryotic mRNA. The poly-A tail can be used to separate eukaryotic mRNA from other RNA species with oligo T cellulose.

- **polyandrous**
having many stamens.

- **polyandry**
the state of having more than one male mate at one time. In fertilisation, the fusion of one female and two or more male pronuclei within an egg cell.

- **polycarpellary**
composed of several carpels.

- **polycarpic**
in general, bearing many fruits (i.e., producing fruit many times or indefinitely), in botany, having a gynoecium forming two or more distinct ovaries or carpels.

- **polycarpous**
See **polycarpic**.

- **polycentric**
  a chromosome that shows more than two centromeres.

- **polycistronic**
  referring to prokaryotic messenger RNAs that contain several cistrons within the same mRNA transcript.

- **polycistronic mRNA**
  an mRNA that codes for more than one protein.

- **polycotyledonous embryo(s)**
  embryos having more than two cotyledons or seed leaves (e.g., in pines and conifers.)

- **polycross**
  open pollination of a group of genotypes (generally selected) in isolation from other compatible genotypes in such a way as to promote random mating inter se it is a widely used procedure for intercrossing parents by natural hybridisation.

- **polycross test**
  a progeny test to assess general combining ability from crosses among selected parents, identities can be maintained only for the seed parents, a mixture of pollen is artificially applied to each female parent topcross.

- **Polycystic Kidney Disease (PKD)**
  a group of conditions characterised by fluid filled sacs that slowly develop in both kidneys, eventually resulting in kidney malfunction.

- **poly-da/poly-dt technique**
  a method of inserting DNA into a vector by adding poly-dA to the linearised vector and poly-dT to the DNA of interest. Also feasible with poly-dG and poly-dC.

- **polydactyly**
  more than five fingers and/or toes. Inherited as an autosomal dominant phenotype.

- **polyembryonic**
  See **polyembryony**.

- **polyembryony**
  the condition in which an ovule has more than one embryo like in certain grasses or cereals. In the past, the phenomenon was used for haploid selection among the embryos, which show often different ploidy levels.

- **polyethylene glycol (PEG)**
  a polymeric substance of molecular weight between 1,000-6,000. It is used for stimulation of protoplast fusion.

- **polygamous**
plants that show male, female and hermaphrodite flowers on the same or different plants.

- **polygenes**
one of a group of genes that together controls a quantitative character. Individually each gene has little effect on the resulting phenotype, which instead requires the interaction of many genes.

- **polygenic**
of traits determined by many genes, each having only a slight effect on the expression of the trait.

- **polygeny**
a trait that is controlled by many genes.

- **polygynoecial**
having a number of pistils joined together, as in aggregate fruits (e.g., raspberry).

- **polyhaploid**
haploid plant derived from a polyploid individual.

- **polyhybrid**
individuals that are heterozygous with respect to the alleles of many gene loci or of crosses involving parents that differ with respect to the alleles of more loci.

- **polykaryotic**
cells showing many nuclei.

- **polylinker**
synthetic oligonucleotide with recognition sites for several restriction endonucleases.

- **polymerase**
any enzyme that catalyses the formation of DNA or RNA from deoxyribonucleotides or ribonucleotides.

- **Polymerase Chain Reaction (PCR)**
a powerful method for amplifying specific DNA segments which exploits certain features of DNA replication. For instance replication requires a primer and specificity is determined by the sequence and size of the primer. The method amplifies specific DNA segments by cycles of template denaturation, primer addition, primer annealing and replication using thermostable DNA polymerase. The degree of amplification achieved is set at a theoretical maximum of 2 ^ N, where N is the number of cycles,e.g., 20 cycles gives a theoretical 1048576 fold amplification.

- **polymerase, DNA or RNA**
enzymes that catalyse the synthesis of nucleic acids on preexisting nucleic acid templates,

assembling RNA from ribonucleotides or DNA from deoxyribonucleotides.

- **polymeric genes**
genes with equal effects but cumulative action.

- **polymerise**
to form a complex compound (polymer) by linking together many smaller elements (residues).

- **polymery**
the production of a trait by cooperation among several polymeric genes.

- **polymix breeding**
an alternative method to full-sib crossing and testing the alternative solution is to apply a pollen mix of many male parents rather than a single pollen for each cross. Polymix breeding is easy to implement, ensures good estimates of breeding values of the parent being pollinated and provides for increased genetic gain opportunity because of the significantly increased number of effective parental combinations tested. However, it has found limited use because inbreeding and pedigree control is lost.

- **polymorphic**
occurring in several different forms.

- **polymorphism**
the occurrence in a population (or among populations) of several phenotypic forms associated with alleles of one gene or homologs of one chromosome.

- **polynemic chromosome**
describes metaphase chromosome and/or chromatids with more than two DNA helices.

- **polynucleotide**
a sequence of many nucleotides.

- **polypeptide**
a chain of linked amino acids, a protein.

- **Polyphenol Oxidase (PPO)**
it is one of the major enzymes responsible for browning of wheat food products wheat cultivars differ in PPO activity and plant breeders wish to select germplasm and cultivars with low PPO activities.

- **polypheny**
pleiotropy.

- **polyphylesis**
originating from several lines of descent.

- **polyphyletic**
designating a group of species arbitrarily classified together, some of the members of which have distinct evolutionary his-

tories, not being descended from a common ancestor.

- **polyploid**
  a cell or an organism having three or more chromosome sets.

- **polyploidisation**
  the spontaneous or induced multiplication of a haploid or diploid genome of a cell or individual.

- **polyploidy**
  the condition in which an individual possesses one or more sets of homologous chromosomes in excess of the normal two sets found in diploids. It can be produced in nature by somatic doubling due to irregular mitosis in the meristematic cell and by unreduced gametes due to irregular reductional division during meiosis, for example, approximately 70 percent of grass species and 20 percent of legumes are polyploid.

- **polyribosome**
  see **polysome**.

- **polysaccharide**
  a biological polymer composed of sugar subunits for example, starch or cellulose.

- **polysaccharide**
  a molecule composed of chains of sugar units.

- **polysome**
  the configuration of several ribosomes simultaneously translating the same mRNA. Shortened form of the term polyribosome.

- **polytene chromosome**
  a giant chromosome produced by an endomitotic process in which, following synapsis of the two homologues, multiple rounds of replication produce chromatids that remain synapsed together in a haploid number of chromosomes. Large chromosome consisting of many chromatids formed by rounds of endomitosis following synapsis of the two homologues.

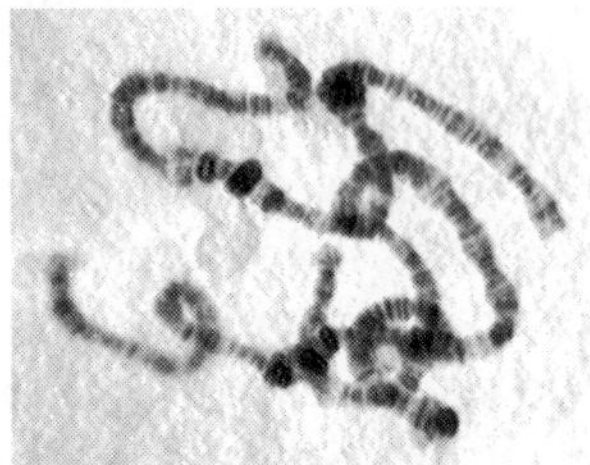

- **polytene chromosome**
  a chromosome that is formed by repeated reduplication of single chromatids. Sections may appear to puff or swell due to differential gene activation. It is visible through the light microscope.

- **polyteny**
  polytene chromosome.

- **pomaceous fruit**
  pome.
- **pome**
  a fruit (e.g., apple, pear, quince) in which the seeds are protected by a tough carpel wall and the entire fruit is embedded in a fleshy receptacle.
- **pomology**
  the science or study of growing fruit.
- **population**
  a group of organisms of the same species relatively isolated from other groups of the same species.
- **population genetics**
  the study of inherited variation in populations and its modulation in time and space. It relates the heri changes in populations to the underlying individual processes of inheritance and development.
- **population size**
  the number of individuals in a population that are included in reproduction during a certain generation.
- **population waves**
  irregular or rhythmic changes of the number of individuals in a population that are included in reproduction during certain generations.
- **population, closed**
  a group of interbreeding plants (occurring in a certain area or in an experimental design) or a group of plants originating from one or more common ancestors, where there is no immigration of plants or pollen.
- **population, Mendelian**
  a group of (potentially) interbreeding plants (cross-fertilising crops), which may occur in a certain area or in an experimental design.
- **porosity**
  the proportion, as the percentage volume, of the total bulk volume of a body of rock or soil occupied by pore space.
- **position (positional) effect**
  the change in the expression of a gene with respect to neighbouring genes.
- **positional cloning**
  a process of molecular cloning of a gene with reference to its position on the genetic or physical map.
- **position-effect variegation**
  variegation caused by the inactivation of a gene in some cells through its abnormal juxtaposition with heterochromatin.

- **positive assortative mating**
  a situation in which like phenotypes mate more commonly than expected by chance.

- **positive control**
  regulation mediated by a protein that is required for the activation of a transcription unit.

- **positive interference**
  when the occurrence of one crossover reduces the probability that a second will occur in the same region.

- **postemergence herbicide**
  a herbicide that affects the weeds after emergence.

- **postgamous incompatibility**
  crossbreeding barrier.

- **postreplicative repair**
  a DNA repair process initiated when DNA polymerase bypasses a damaged area. enzymes in the rec system are used.

- **Post-Transcriptional Gene Silencing (PTGS)**
  silencing of an endogenous gene caused by the introduction of a homologous dsRNA, transgene or virus. In PTGS, the transcript of the silenced gene is synthesised but does not accumulate because it is rapidly degraded. This is a more general term than RNAi, since it can be triggered by several different means. It was initially considered a bizarre phenomenon limited to petunias and a few other plant species. In the 1990s, a surprising observation was made in petunias. While trying to deepen the purple colour of these flowers, R. Jorgensen introduced a pigment-producing gene under the control of a powerful promoter. Instead of the expected deep purple colour, many of the flowers appeared variegated or even white. The observed phenomenon was named 'cosuppression since the expression of both the introduced gene and the homologous endogenous gene was suppressed, meanwhile, it became one of the most important phenomenon in molecular biology. It has become clear that PTGS occurs in both plants and animals and has roles in viral defence and transposon silencing mechanisms. Most exciting is the emerging use of PTGS and, in particular, RNA interference (RNAi)PTGS initiated by the introduction of double-stranded RNA (dsRNA)as a tool to knock out expression of specific genes.

- **post-transcriptional modification**
  changes in eukaryotic mRNA, tRNA or other RNAs made after

transcription has been completed. The changes to mRNA include addition of a 5'cap and 3' polyA tail and removal of introns.

- **postzygotic incompatibility**
  a condition where, in the case of incompatible or wide crosses, the zygote fails to develop, often for nutritional reasons. In some cases embryo culture can be used to rescue the embryo.

- **potassium (K)**
  an element that is required for plant growth, deficiency leads to reduced growth and to dark or blue-green coloration in the leaves.

- **pot-bound**
  used when a plant has an overly extensive root system in a too-small container.

- **power (of statistical test)**
  it is the probability that a statistical test will detect a defined pattern in data and declare the extent of the pattern as showing statistical significance.

- **PPRP**
  Plant Pathogenesis-Related Proteins.

- **preadaptation**
  an adaptation evolved in one adaptive zone, which proves especially advantageous in an adjacent zone and so allows the organism to radiate into it. In breeding, the pretreatment of plants under moderate climatic (light, temperature or nutrient) conditions in order to gradually accustom them to stress conditions.

- **prebasic seed**
  breeder('s) seed.

- **prebloom**
  the stage or period immediately preceding blooming.

- **prebreeding**
  all research and screening activities before a plant material enters the directed breeding process (e.g., the development of germplasm to a state where it is viable for breeder's use). Primarily, it involves the evaluation of traits from exotic material and their introduction into more cultivated backgrounds.

- **prechilling**
  the practice of exposing imbibed seeds to cool (+5 to +10°C) temperature conditions for a few days prior to germination at warmer conditions.

- **precipitant**
  a substance or process that causes precipitation.

- **precision seed**
calibrated or pelleted sorts of seeds used for precision drilling.

- **precleaning**
the process for removing the bulk of foreign materials grossly different in size from the harvested seed or other crop products.

- **precocious embryo development**
asexual development of the embryo before the flower opens and anthers dehisce.

- **predecessor**
within the crop rotation, the crop before the recent cropping.

- **predisposition**
an increase in susceptibility resulting from the influence of environment on the suspect.

- **preferential pairing**
chromosome pairing in allopolyploids in which the most structurally similar chromosomes preferentially pair with another *Ph* locus.

- **pre-mRNA**
the first (primary) transcript from a protein coding gene is often called a pre-mRNA and contains both introns and exons. Pre-mRNA requires splicing (removal) of introns to produce the final mRNA molecule containing only exons.

- **preselection**
selection of some individuals for further tests or studies.

- **presoaking**
presoaking of seeds in water has been suggested as a means to speed up germination.

- **presymptomatic diagnosis**
diagnosis of a genetic condition before the appearance of symptoms.

- **prezygotic incompatibility**
in the case of incompatible or wide crosses the inhibition of pollen germination or the prevention of pollen tube growth, among other possible barriers of plant fertilisation. In some cases the barrier can be overcome by in vitro pollination.

- **pribnow box**
relatively invariant sequence of six nucleotides (consensus TATAAT) in prokaryotic promoters centred at the position minus 10 (10 base pairs upstream from the transcription start site).

- **pricking off**
a method of transplanting tiny seedlings, the blade of a knife or plant marker is used to remove each plant from one spot and move it to another.

■ **pricking out**
a method of thinning seedlings by cutting them off at soil level so as not to disturb the roots of the other plants.

■ **pricking-out peg**
an adjusted peg used for thinning or transplanting seedlings by cutting them off at soil level.

■ **prickly**
having thorns.

■ **primary culture**
a culture resulting from cells, tissues or organs taken from an organism.

■ **primary fluorescence**
fluorescence originating from the specimen itself.

■ **primary gene pool**
it includes the cultivated species of a crop and related species from which useful genes can be most readily obtained for breeding. In general, it is the total sum of all the genetic variation in the breeding population of a species and closely related species that commonly interbreed with or can be routinely crossed with, the species.

■ **primary infection**
the first infection of a plant by the overwintering or oversummering pathogen.

■ **primary inoculum**
spores or fragments of a mycelium capable of initiating a disease.

■ **primary leaf**
one of the first pair of leaves to emerge above the cotyledon during the development of a seedling, it is often morphologically distinct from subsequent leaves.

■ **primary structure**
the sequence of polymerised amino acids in a protein.

■ **primary transcript**
the product of eukaryotic transcription before post-transcriptional modifications take place.

■ **primase**
an enzyme that creates an RNA primer for initiation of DNA replication.

■ **primer**
a short sequence (of RNA or DNA) from which DNA replication can initiate. May be either a synthetic DNA or RNA or a length of RNA synthesised in vivo by primase.

■ **priming**
the treatment of seeds with an osmotic solution (e.g., polyethylene glycol, which allows controlled hydration), the seed embryo develops to the point

of germination and then is dried. It is applied for more uniform and rapid germination of certain vegetable seeds.

- **primitive form**
in phylogeny, seedless vascular plants with underground rhizomes. They grow in tropical to subtropical areas and are terrestrial or epiphytic. In botany and breeding, as compared to cultivated crops, plants still show wild characteristics, such as brittle rachis, seed shattering and others.

- **primordium**
the early cells that serve as the precursors of an organ to which they later give rise by mitotic development.

- **primosome**
a complex of two proteins, a primase and a helicase, that initiates RNA primers on the lagging DNA strand during DNA replication.

- **principal host**
host.

- **pro**
proline (an amino acid).

- **probability**
the long term frequency of an event relative to all alternative events and usually expressed as decimal fraction.

- **probability theory**
the conceptual framework concerned with quantification of probabilities.

- **probe**
1. single- stranded DNA or RNA molecules of specific base *sequence*, labelled either radioactively or immunologically, that are used to detect the complementary base sequence by hybridisation.
2. a radioactively or nonradioactively labelled and defined nucleic acid sequence that can be used in molecular hybridisation. Usually it is used for ex situ or in situ identification of specific, complementary nucleic acid sequences.

- **processing of seeds**
the complex of measures in order to clean, calibrate, disinfect, store and pack seeds.

- **processivity**
the ability of an enzyme to repetitively continue its catalytic function without dissociating from its substrate.

- **prochromosome**
a heterochromatin block that is seen during the interphase of cell division and which is related to the number of chro-

mosomes per complement or less.

- **procumbent**
trailing or laying flat on the ground.

- **product of meiosis**
one of the (usually four) cells formed by the two meiotic divisions.

- **proflavin**
a mutagen that tends to produce frameshift mutations.

- **progenitor cell**
undifferentiated cell (i.e., immature cell) which will go on to develop into any cell type.

- **progeny**
offspring, plants grown from the seeds produced by parental plants.

- **progeny selection**
selection based on progeny performance.

- **progeny testing**
breeding of offspring to determine their genotypes and that of their parents.

- **prognosis**
prediction of the course and probable outcome of a disease.

- **prokaryote**
an organism lacking a true nucleus, such as a bacterium or a blue-green alga.

- **prokaryotic cell**
a cell having no nuclear membrane and hence no separate nucleus.

- **prolamin**
a protein, it is soluble in 70-90 percent ethyl alcohol but not in water. It is found only in cereal seeds (e.g., gliadins in wheat and rye, zein in maize), upon hydrolysis they yield proline, glutamic acid and ammonia.

- **proliferation**
successive development of new parts, organs, etc.

- **prometaphase**
the stage in mitosis between the dissolution of the nuclear membrane and the organisation of the chromosomes on the metaphase plate.

- **promiscuous**
heterogeneous or haphazard mixture.

- **promoter**
1. a regulatory region a short distance upstream from the 5' end of a transcription start site that acts as the binding site for RNA polymerase. A region of DNA to which RNA polymerase binds in order to initiate transcription. 2. a nucleotide sequence within an operon, lying between the operator and the structural

gene or genes, that serves as a recognition site and point of attachment for the RNA polymerase. It is the starting point for transcription of the structural gene(s) in the operon, but is not itself transcribed. The promoter controls where (e.g., which portion of a plant) and when (e.g., which stage in the lifetime) the gene is expressed. For example, the promoter Bce4 is seed-specific cauliflower mosaic virus (CaMV).

- **promoter sequence**
  promoter.

- **proofread**
  DNA polymerase has 3' to 5' exonuclease activity which is used during polymerisation to remove nucleotides it has recently added if they are incorrectly paired and is a correcting ability to remove errors in replication.

- **propagate**
  to reproduce or cause to multiply or breed.

- **propagating bench**
  a stationary, shallow box (sometimes covered by a glass pan or other means), it is usually filled with fine sand or certain soil (often sterilised), which is kept moist. Cuttings, slips or shoots after in vitro culture are inserted into the growing medium until they form roots.

- **propagation**
  various methods by which plants are increased (e.g., seeds, division, separation, softwood cuttings, slips, grafting, budding or layering).

- **propagule**
  any type of plant to be used for reproduction (e.g., seedling, a rooted or unrooted cutting, a graft, a tissue-cultured plantlet, etc.).

- **prophage**
  a phage chromosome inserted as part of the linear structure of the DNA chromosome of a bacterium. A temperate phage integrated into the host chromosome.

- **prophyll(um)**
  the first leaf or protective scale of a lateral shoot.

- **proplastid**
  1. mutant plastids that do not grow and develop into chloroplasts.
  2. a colourless, double-membrane-bound organelle with little internal structure that acts as a precursor in the development of all plastids.

- **protandry**
  the maturation of anthers before carpels (e.g., in sugarbeet, sunflower or carrot).

- **protease**
  an enzyme that catalyses the hydrolysis of the peptide bonds of proteins and peptides proteinase.

- **protected variety**
  a variety that is released and granted a certificate of plant variety protection under the legal statutes of a given country. The owner of a protected variety has the right of selling, offering, reproduction, import, export or using for hybrid seed production.

- **protein**
  a large molecule composed of one or more chains of amino acids in a specific order. The order is determined by the base sequence of nucleotides in the coding for the protein. Proteins are required for the structure, function and regulation of the body's cells, tissues and organs and each protein has unique functions. Examples are hormones, enzymes and antibodies.

- **protein engineering**
  production of altered proteins by site-directed mutagenesis biotechnology.

- **proteinase**
  an enzyme that hydrolyses protein molecules.

- **proteome**
  the complete set of proteins detec in a tissue.

- **proteome analysis**
  the basis of proteome analysis is an electrophoretic separation of the proteins on a two-dimensional protein gel, silver staining of proteins, followed by an image analysis of the stained gel. Interesting protein spots identified on the gel can be excised, the critical protein can be extracted from the excised spot and further analysis of the protein (amino acid composition, partial amino acid sequence, isoelectric point, molecular weight) may result in protein identification.

- **proteomics**
  systematic characterisation of proteins, which are found in a tissue or in a specific physiological condition. The proteins can be identified using mass spectroscopy.

- **protoclonal variation**
  variability of somatic cells derived from protoplast culture.

- **protoclone**
a plant regenerated from a protoplast culture.

- **protogyny**
a condition in which the female parts develop first (e.g., in rapeseed).

- **proto-oncogene**
the non-activated form of a cellular oncogene in an untrans-formed cell. A gene that, when mutated or otherwise affected, becomes an oncogene.

- **protoplasm**
a complex, translucent, colou-rless, colloidal substance within each cell, including the cell membrane, but excluding the large vacuoles, masses of secretions, ingested material, etc.

- **protoplast**
a plant cell whose wall has been removed.

- **protoplast culture**
the isolation and culture of plant protoplasts by mechanical means or by enzymatic digestion of plant tissue, organs or cultures derived from these. Protoplasts are utilised for selection or hybridisation at the cellular level and for a variety of other purposes.

- **protoplast fusion**
a technique used in somatic hybridisation experiments, it is used for overcoming crossing barriers. Protoplasts are placed together and induced to fuse, applying fusogenic agents, such as polyethylene glycol or physical means. Subsequent regeneration of the cell wall allows the propagation and regeneration of a somatic hybrid plant.

- **prototroph**
a strain of organisms capable of growth on a defined minimal medium from which they can synthesise all of the more complex biological molecules they require, as opposed to auxotroph.

- **provenance**
the geographical source and/or place of origin of a given lot of seed, propagules or pollen.

- **provenance test**
an experiment, usually replicated, comparing trees grown from seed or cuttings that were collected from many geographical regions of a species distribution.

- **provirus**
a virus chromosome integrated into the DNA of the host cell.

- **proximal**
  toward or nearer to the place of attachment.

- **pruning**
  trimming branches or parts of trees and shrubs in order to trim a plant or to bring the plant into a desired shape, in addition, the removal of the growing point from a plant frequently causes the initiation of tillering or branching. The onset of flowering on the new vegetative growth may be delayed compared with that of unpruned plants, the method may be also used for synchronisation of flowering prior to hybridisation.

- **pseudoallele**
  allele that is functionally but not structurally allelic, that is wild-type recombinants can be recovered by intragenic recombination from heterozygotes containing two different pseudoalleles.

- **pseudoautosomal gene**
  a gene that occurs on both sex-determining heteromorphic chromosomes.

- **pseudobivalent**
  a bivalent-like association of two mitotic chromosomes due to reciprocal chromatid exchange.

- **pseudocarp**
  a fruit consisting of one or more ripened ovules attached or fused to modified bracts or other nonfloral structures.

- **pseudocompatibility**
  the occurrence of fertilisation that normally is prevented by incompatibility mechanisms. It is caused by specific environmental or genotypic conditions, for example, in rye, by high temperature (about +30°C) self-incompatibility can be broken so that seeds are set.

- **pseudodominance**
  the sudden appearance of a recessive phenotype in a pedigree, due to deletion of a masking dominant gene. The phenomenon in which a recessive allele shows itself in the phenotype when only one copy of the allele is present, as in hemizygous alleles or in deletion heterozygotes.

- **pseudogamous heterosis**
  increased vigour of maternal offspring due to male parent influence on the endosperm.

- **pseudogamy**
  a type of apomixis in which the diploid egg cell develops into the embryo without fertilisation of the egg cell, although only after fertilisation of the polar

nuclei with one of the sperm cells from the male gamete to form a normal triploid endosperm.

- **pseudogene**
  an inactive gene derived from an ancestral active gene.

- **pseudoheterosis**
  it designates hybrids between species, varieties or lines that exceed the parents in some traits.

- **pseudoisochromosome**
  a chromosome that shows only equal ends as a result of interchanges.

- **pseudovivipary**
  vegetative proliferation of plantlets in the inflorescence axes.

- **pto**
  a specific resistance gene in tomato, which controls resistance against the bacterium *Pseudomonas syringae* pv. *tomato*. The gene encodes a cytosolic protein kinase. The Pto system remains the best characterised specific resistance gene system.

- **pubescent**
  covered with soft hairs, downy.

- **puff**
  a structural modified region of a polytene chromosome, it originates from the despiralisation of deoxyribonucleoprotein.

- **pulp**
  the soft, succulent part of a fruit, usually composed of mesocarp and/or the pith of a stem.

- **pulse**
  the edible seeds of any leguminous crop legume(s).

- **pulse-chase experiment**
  an experiment in which cells are grown in radioactive medium for a brief period (the pulse) and then transferred to nonradioactive medium for a longer period (the chase).

- **pulsed-field gel electrophoresis**
  an electrophoretic technique in which the gel is subjected to electrical fields alternating between different angles, allowing very large DNA fragments to snake through the gel and hence permitting efficient separation of mixtures of such large fragments.

- **punctuated equilibrium**
  the evolutionary process involving long periods without change (stasis) punctuated by short periods of rapid speciation.

- **punnett square**
  a diagrammatic representation of a particular cross used to pre-

dict the progeny of the cross. A grid is used as a graphic representation of the progeny zygotes resulting from different gamete fusions in a specific cross.

- **pure bred**

derived from a line subjected to inbreeding.

- **pure line**

a number of individuals of a successive, self-pollinated crop, which derives from a single plant, a strain homozygous at all loci.

- **pure line breeding**

true breeding.

- **pure-breeding line or strain**

a group of identical individuals that always produce offspring of the same phenotype when intercrossed.

- **pure-live seed**

the percentage of the content of a seed lot that is pure and viable. It is determined by multiplying the percentage of pure seed by the percentage of viable seed (germination percentage) and dividing by 100.

- **purine**

a nitrogen- containing, single-ring, basic compound that occurs in nucleic acids. The purines in DNA and RNA are adenine and guanine.

- **purity testing**

determination of the degree of contamination of seed lots with genetically nonidentical, damaged or pest-infected seeds.

- **puroindoline proteins**

small, basic, cysteine-rich proteins found in bread wheat (*Triticum aestivum*). By engineered introduction of a puroindoline gene the hard textured grain of durum wheat (*Triticum durum*) and other cereals, including maize and barley, can be converted into soft texture, puroindoline genes. Puroindoline a and puroindoline b, represent the grain hardness gene on chromosome 5DS, in hexaploid wheat, soft kernel texture is the wild type, whereas hard texture (hardness) results from any one of several mutations in the puroindolines.

- **pycnidium**

a flask-shaped fungal receptacle bearing asexual spores (i.e., pycniospores).

- **pycniospore**

a spore from a pycnidium.

- **pycnotic**

the concentration of the nucleus into a compact, strongly stained mass, taking place as the cell dies.

- **pyrenoid**
it refers to a proteinaceous region of chloroplast, sometimes the site of accumulation of carbohydrate reserves.

- **pyriform**
pear-shaped.

- **q banding**
a chromosomal staining technique using the fluorescence dye quinacrin mustard. Under UV light a characteristic light and dark banding is induced.

- **quadrivalent**
a chromosome association of four members.

- **qualitative character**
a character in which variation is discontinuous.

- **qualitative inheritance**
an inheritance of a character that differs markedly in its expression amongst individuals of a species. Variation is discontinuous. Such characters are usually under the control of major genes.

- **qualitative resistance**
vertical resistance.

- **quantitative character**
a character in which variation is continuous so that classification into discrete categories is arbitrary; metric character.

- **quantitative genetics**
a branch of genetics that deals with the inheritance of quantitative traits, sometimes it is also called biometrical or statistical genetics.

- **quantitative inheritance**
the mechanism of genetic control of traits showing continuous variation.

- **quantitative resistance**
horizontal resistance.

- **quantum speciation**
the rapid rise of a new species, usually in small isolates, with the 'founder effect' and random genetic drift. It is also called saltational speciation.

- **quarantine**
the official confinement of plants subject to phytosanitary regulations for observation and research or for further inspection and/or testing. More general, a legal ban on the export or import of certain noxious weeds or insects that may be attached to the plants.

- **quarternary hybrid**
a hybrid derived from four different grandparental individuals.

- **quartet**
the four nuclei and/or cells produced during meiosis tetrad.

- **quaternary structure**
the association of polypeptide subunits to form the final structure of a protein.

- **quelling**
post-transcriptional gene silencing in *Neurospora crassa* induced by the introduction of a transgene PTGS.

- **quereitron**
glucoside.

- **quick test (of seed testing)**
a type of test for evaluating seed quality, usually germination, more rapidly than standard laboratory tests.

- **quiescence**
dormancy.

- **quincunx planting**
planting four young plants to form the corners of a square with a fifth plant at its centre.

- **quinone**
growth inhibitor.

- **r plasmid**
a plasmid containing one or several transposons with resistance genes. Plasmids carrying genes control resistance to various drugs.

- **r1, r2, r3, etc.**
the first, second, third, etc. generation following any type of irradiation in mutation breeding.

- **rabl configuration**
in plants and insects, it refers to chromosomes that are spatially organised within the nucleus, with centromeres clustered on the nuclear membrane at one pole and telomeres attached to the nuclear membrane of the other hemisphere.

- **race**
a genetically and, as a general rule, geographically distinct interbreeding division of a species. In other contexts, a population or group of populations distinguishable from other such populations of the same species by the frequencies of genes, chromosomal rearrangements or hereditary phenotypic characteristics. A race that has received a taxonomic name is a subspecies.

- **raceme**
an inflorescence in which the main axis continues to grow, producing flowers laterally, such that the youngest ones are apical or at the centre.

- **racemose**
an indeterminate, unbranched inflorescence in which the flowers are borne on pedicels of about equal length, along an elongated axis.

- **race-nonspecific type**

host-plant resistance that is operational against all races of a pathogen species. It is variable, sensitive to environmental changes and usually polygenetically controlled.

- **race-specific resistance**

it is (individually) determined by single or very few genes in the plant and these are most usually effective against a limited proportion of the pathogen population. Synonyms: vertical and qualitative resistance. The proportion of the pathogen population capable of causing disease on a resistant host is termed virulent on the host. Correspondingly, the host is susceptible or resistant and the interaction compatible or incompatible, respectively, where the inheritance of virulence/avirulence has been studied. It is generally found that a specific avirulence gene in the pathogen corresponds to specific resistance gene in the plant. This relationship is termed the gene-for-gene-hypothesis.

- **race-specific type**

host-plant resistance that is operational against one or a few races of a pathogen species, generally produces an immune or hypersensitive reaction and is controlled by one or a few genes.

- **rachilla**

the spikelet axis, it is also applied to the segment of the rachilla that remains attached to the oat grain.

- **rachis**

the main axis of the ear and of the panicle of grasses.

- **RAD**

an abbreviation for 'Radiation Absorbed Dose,' a measure of the amount of any ionising radiation that is absorbed by the tissue. One RAD is equivalent to 100 ergs of energy absorbed per gram tissue.

- **radicle**

the root of the embryo, which develops into the primary root of the seedling.

- **radioactive**

a substance when a constituent chemical element is undergoing the process of change into another element through the emission of radiant energy. Radioactivity is used as a tool in research to tag or trace the movement of compounds. The presence of a compound containing a radioactive element is revealed by instruments that measure the radiant energy emitted or by radiosensitive films .

radioactive tracer
isotopic tracer.

radiocarbon dating
a dating method for organic material that is applicable to about the last 70,000 years. It relies on the assumed constancy over time of atmospheric ^{14}C:^{12}C ratios and the known rate of decay of radioactive carbon, of which half is lost in a period of about every 5.730 years.

raffinose
the raffinose family of oligosaccharides, is believed to play an important role in the resistance of plants to environmental stress by protecting membrane-bound proteins. It is relevant when seeds mature because they rapidly lose moisture as they dry out.

rain forest
a tropical forest of tall, densely growing, broad-leaved evergreen trees in an area of high annual rainfall.

ram mutation
ribosomal ambiguity mutation that allows incorrect tRNAs to be incorporated into the translation process.

ramet
an individual that belongs to a clone.

ramification
the act or process of ramifying (i.e., the repeated division of branches into secondary branches).

ramify
ramification.

random effects
effects of a treatment in an experiment in which the treatments are considered in relation to a whole range of possible treatment effects.

random genetic drift
changes in allelic frequency due to sampling error. Changes in allele frequency that result because the genes appearing in offspring are not a perfectly representative sampling of the parental genes. (e.g., in small populations.)

random lines
isogenic lines.

random mating
mating between individuals where the choice of partner is not influenced by the genotypes (with respect to specific genes under study.) The mating of individuals in a population such that the union of individuals with the trait under study occurs according to the product rule of probability.

- **random sampling**
  a sample drawn from a population in such a way that every individual of the population has an equal chance of appearing in the sample. It ensures that the sample is representative and provides the necessary basis for virtually all forms of inference from sample to population, including the informal inference, which is characteristic of rerandomisation statistics.

- **random strand analysis**
  mapping studies in organisms that do not keep together all the products of meiosis.

- **randomisation**
  the process of making assignments at random. In field trials, randomisation of entries is required to obtain a valid estimate of experimental error. Each entry must have an equal chance of being assigned to any plot in a replication and an independent randomisation is required for each replication.

- **randomised block**
  the entries to be tested are assigned at random to the plots within the block.

- **randomised block design**
  a randomised block analysis of variance design (e.g., one-way blocked ANOVA) is created by first grouping the experimental subjects into blocks, the subjects in each block are as similar as possible (e.g., littermates), there are as many subjects in each block as there are levels of the factor of interest. Randomly assigning a different level of the factor to each member of the block, such that each level occurs once and only once per block, the blocks are assumed not to interact with the factor randomised.

- **Randomised Complete Block (RCB)**
  each block contains a plot for each candidate to be tested. In this case the classification of the data according to the blocks and the classification according the candidates are orthogonal.

- **range of reaction**
  the range of all possible phenotypes that may develop by interaction with various environments from a given genotype, it is also called 'norm of reaction'.

- **range pole**
  landmark.

- **RAPD**
  Random Amplified Polymorphic DNA technique.

- **raphanobrassica**

an intergeneric hybrid between *Raphanus* and *Brassica* species. A first hybrid was already reported in 1826, about 100 years later Karpechenko produced a hybrid between *Raphanus sativus* ($2n = 2x = 18$, RR) x *Brassica oleracea* ($2n = 2x = 18$, CC) $2n = 4x = 36$, RRCC.

- **raphe**

a ridge, sometimes visible on the seed surface, which is the axis along which the ovule stalk joins the ovule.

- **rate of response to selection**

selection progress.

- **ratooning**

a sprout or shoot from the root of a plant (e.g., in sugarcane) after it has been cropped.

- **ray**

a pedicel in an umbellate inflorescence.

- **ray flower**

a flower head with outer ray flowers forming petallike structures surrounding the inner disc flowers (e.g., in the *Asteraceae* or in sunflower).

- **rDNA**

ribosomal DNA.

- **reading frame**

the mechanism that moves a ribosome, one codon at a time, from a designated start sequence during genetic translation, a shift in the reading frame by any number of nucleotides other than three or multiples of three will cause an entirely new sequence of codons.

- **reading mistake**

the placement of an incorrect amino acid into a polypeptide chain during protein synthesis.

- **readthrough**

transcription or translation beyond the normal termination signals in DNA or mRNA respectively.

- **reagent**

a substance that, because of the reactions it causes, is used in analysis and synthesis.

- **realised genetic gain**

it refers to the observed difference between the mean phenotypic value of the offspring of the selected parents and the phenotypic value of the parental generation before selection.

- **realised heritability**

Heritability measured by a response to selection. The ratio of the single-generation progress of selection to the selection differential of the parents.

- **reannealing**
  spontaneous realignment of two single DNA strands to re-form a DNA double helix.

- **reaper**
  a machine for cutting standing grain.

- **reaper-binder**
  an implement that cuts and ties hay into bundles.

- **rearrangement**
  all chromosome mutations that result in modified karyotypes.

- **rec system**
  several loci controlling genes (recA, recB, recC and others) involved in postreplicative DNA repair.

- **recalcitrant seed**
  seed that does not survive drying and freezing, in particular, seed that cannot withstand either drying or temperatures of less than +10°C and, therefore, cannot be stored for long periods, as compared to orthodox seeds.

- **receptacle**
  the part of the stem from which all parts of the flower arise. In compositae, the flattened tip of the stem that bears the bracts and florets.

- **receptaculum**
  receptacle.

- **receptor element**
  a controlling element in maize that can insert into a gene (making it mutant) and that can also excise (thus making the mutation unstable) both of these functions are nonautonomous, being under the influence of a regulator element.

- **receptor site**
  in molecular genetics, a set of reactive chemical groups in the cell wall of a bacterium that are complementary to a similar set in the tailpiece of bacteriophage.

- **recessive**
  an allele that is not expressed in the heterozygous condition. Also the phenotype of the homozygote of a recessive allele.

- **recessive epistasis**
  the effect of a recessive allele of a gene suppressing the phenotype manifestation of another gene.

- **recessive phenotype**
  the phenotype of a homozygote for the recessive allele, the parental phenotype that is not expressed in a heterozygote.

- **recessiveness**
recessive.

- **recipient**
one that receives, receiver.

- **recipient cell**
recipient.

- **reciprocal altruism**
an apparently altruistic behaviour performed with the understanding that the recipient will reciprocate at some future date.

- **reciprocal cross**
a cross, with the phenotype of each sex reversed as compared with the original cross, to test the role of parental sex on inheritance pattern. A pair of crosses of the type genotype A(female) X genotype B(male) and genotype B(female) X genotype A(male).

- **reciprocal full-sib selection**
a method of interpopulation improvement for species in which the commercial product is hybrid seed. A cycle of selection is completed in the fewest number of seasons by use of plants from which both selfed and hybrid seed can be obtained.

- **reciprocal genes**
nonallelic genes that reciprocate or complement one another.

- **reciprocal half-sib selection**
reciprocal recurrent selection.

- **reciprocal recurrent selection**
a breeding method used to achieve an accumulation of genes that are valuable for specific traits but also for combining ability. In practice, two populations form the basis of selection. Reciprocally, one population serves as a tester for the investigation of the selections deriving from the other population. From the first population, usually a greater number of plants is selfed and at the same time crossed with several plants of the second population. The same procedure is realised with the second population. During the second year, the progenies of the test-crosses are subjected to performance trials. The progenies of the cross of one female parent, derived from population 1, are combined with several male parents from population 2. Based on the performance testing, progenies from selfed seed, produced by best plants, are grown during the third year. In the same year, the best progenies of selfings are crossed in many combinations Seeds obtained from those crosses form the improved

populations for growing during the fourth year. Within such populations individual plants may be selfed again and crossed for utilisation in a new cycle recurrent selection.

- **reciprocal translocation**
  a chromosomal configuration in which (usually) the ends of two non-homologous chromosomes have become exchanged. A translocation in which part of one chromosome is exchanged with a part of a separate non-homologous chromosome.

- **recognition sequence (site)**
  a nucleotide sequence composed typically of 4, 6 or 8 nucleotides, it is recognised by a restriction endonuclease, so-called type II enzymes cut (and their corresponding modification enzymes methylate) within or very near the recognition sequence.

- **recombinant**
  in genetic mapping studies an offspring having a non-parental allele combination. An individual or cell with a genotype produced by recombination.

- **recombinant DNA**
  a novel DNA sequence formed by the joining, usually in vitro, of two non-homologous DNA molecules.

- **recombinant DNA technologies**
  procedures used to join together DNA segments in a cell-free system (an environment outside a cell or organism). Under appropriate conditions, a recombinant DNA molecule can enter a cell and replicate there, either autonomously or after it has become integrated into a cellular chromosome.

- **recombinant protein**
  a protein synthesised from a cloned gene.

- **recombinant type**
  an association of genetic markers found among the progeny of a cross that is different from any association of markers present in the parents.

- **recombination**
  the process by which progeny derive a combination of *genes* different from that of either parent. In higher organisms, this can occur by *crossing over*.

- **recombination frequency**
  the number of recombinants divided by the total number of progeny, expressed as a percentage or fraction. Such frequencies indicate relative distances between loci on a genetic map mapping.

■ **recombination nodule**
swellings along DNA strands and associated proteins during prophase pairing of chromosomes. The nodules can be identified under the electron microscope. It is suggested that recombination nodules are the sites of genetic crossing over.

■ **recombination system**
all factors that mediate and control the process of genetic recombination.

■ **recombinational repair**
the repair of a DNA lesion through a process, similar to recombination, that uses recombination enzymes.

■ **recon**
the smallest unit of DNA capable of recombination.

■ **recurrent (backcross) parent**
recurrent parent.

■ **recurrent full-sib selection**
a method of intrapopulation improvement that involves the testing of paired-plant crosses. It is the only method of recurrent selection in which the seeds from two individuals, rather than one, are used for testing and to form the new population.

■ **recurrent parent**
the parent to which a hybrid is crossed in a backcross, it replaces the dragged alleles step by step with the alleles of the original variety.

■ **recurrent reciprocal selection**
a recurrent selection breeding system in which genetically different groups are maintained and, in each selection cycle, individuals are mated from the different groups to test for combining ability recurrent selection.

■ **recurrent selection**
a method designed to concentrate favourable genes scattered among a number of individuals, it is performed by repeated selection in each generation among the progeny produced by matings inter se of the selected individuals of the previous generation. In practice, plants from a population are selfed and, after the yield of the selfed seeds, the progenies of the phenotypically best individuals are grown in the second year. The best progenies are then crossed in as many combinations as possible and the seeds received hereby are grown in the third year as a population. Within the already improved population, selection and selfing can be carried out again. With this population a second cycle of recurrent selection can be started.

■ **rediploidisation**
in anther culture, the haploidisation of the genome by culturing pollen grains to haploid plantlets and its rediploidisation after spontaneous or induced doubling of the chromosome set.

■ **reductase**
an enzyme responsible for reduction in an oxidation-reduction reaction.

■ **reduction division**
a nuclear division that produces daughter nuclei each having one-half as many centromeres as the parental nucleus. The first meiotic division reduces the number of chromosomes and centromeres to half that of the original cell.

■ **redundant DNA**
DNA that does not appear to be genetically active and hence is not translated or transcribed, it often consists of repeated sequences.

■ **redundant gene**
a gene that is present in many functional copies, so that one copy can complement the loss of another copy.

■ **regression**
a term coined by Galton for the tendency of the quantitative traits of offspring to be closer to the population mean than are their parents' traits. It arises from a combination of factors - dominance, gene interactions and environmental influences on traits.

■ **regression coefficient**
the slope of the straight line that most closely relates two correlated variables.

■ **regulator gene**
a gene primarily involved in controlling another (structural) gene. Genes that are involved in turning on or off the transcription of structural genes.

■ **relaxed mutant**
a mutant that does not exhibit the stringent response under amino acid starvation.

■ **repeat sequences**
the length of a nucleotide sequence that is repeated in a tandem cluster.

■ **repetitive DNA**
DNA made up of copies of the same or nearly the same nucleotide sequence. DNA sequences that are present in many copies per chromosome set. Repetitive DNA sequences may be closely linked (e.g., satellite DNA or VNTR loci) or dispersed

throughout the genome or parts of the genome (e.g., alu family.)

- **replica plating**
  a technique to rapidly transfer micro-organism colonies from a master plate, in an exact spatial pattern, to a number of further plates.

- **replication**
  DNA synthesis. The process of copying.

- **replication error**
  any modification that prevents or disturbs the DNA replication process.

- **replication fork**
  the point at which the two strands of DNA are separated to allow replication of each strand.

- **replication unit**
  See **replicon.**

- **replicative transposition**
  the transposition of a transposable element to a new location without it being lost from the original location.

- **replicon**
  a chromosomal region whose replication is controlled by a single adjacent DNA replication initiation site. A genetic unit of replication including a length of DNA and its site for initiation of replication.

- **replisome**
  the DNA-replicating structure at the replication fork consisting of two DNA polymerase III enzymes and a primosome (primase and DNA helicase.)

- **reporter gene**
  a gene whose phenotypic expression is easy to monitor, used to study promoter activity in different tissues or developmental stages. Recombinant DNA constructs are made in which the reporter gene is attached to a promoter region of particular interest and the construct transfected into a cell or organism.

- **repressible operon**
  synthesis of a co-ordinated group of enzymes involved in a single synthetic (anabolic) pathway. Is repressible if excess quantities of (usually) the end product of the pathway leads to cessation of transcription of the genes encoding the enzymes of the pathway.

- **repression**
  the condition of an inactive gene. Lack of activity (transcription) due to presence of repressor.

- **repressor**
  the protein product of a regulator gene that acts to control tran-

scription of inducible and repressible operons. A molecule that binds to the operator and prevents transcription of an operon. Generally any molecule that can reversibly inactivate a gene.

- **reproduce**
to create another individual of the parental type that will in turn produce another reproduction.

- **reproduction**
the process of forming new individuals of a species by sexual or asexual ways.

- **reproductive isolating mechanism**
any environmental, behavioural, mechanical and physiological barrier that prevents two individuals of different populations from producing viable progeny.

- **reproductive isolation**
the absence of interbreeding between members of different species.

- **reproductive meristem**
generative meristem.

- **reproductive organ**
usually, it refers to the sexual organ.

- **reproductive success**
the relative production of offspring by a particular genotype.

- **repulsion**
allelic arrangement of two linked heterozygous loci, in which each homologous chromosome has one mutant (a or b) and one wild-type (A or B) allele (i.e., Ab/aB). Two linked heterozygous gene pairs in the arrangement, Ab/aB.

- **reselection**
backward selection.

- **residue seed method**
method of overstored seeds.

- **resilience**
the ability of a population to persist in a given environment despite disturbance or reduced population size, based upon the ability of individuals within the population to survive (fitness) and reproduce (fecundity) in a changed environment.

- **resin**
an exudate of tree wood or bark, liquid but becoming solid on exposure to air, consisting of a complex of terpenes and similar compounds.

- **resistance**
inherent capacity of a host plant to prevent or retard the development of an infectious disease, there are different types of resistance: (1) hypersensitivity (infection by the pathogen is

prevented by the plant), (2) specific resistance (specific races of the pathogen cannot infect the plant), (3) nonuniform resistance (the host prevents the establishment of certain races), (4) major gene resistance (races of the pathogen are controlled by major genes in the host), (5) vertical resistance (host resistance controls one or a certain number of races), (6) field resistance (severe injury in the laboratory, but resistance under normal field conditions), (7) general resistance (the host is able to resist the development of all races of the pathogen), (8) nonspecific resistance (host resistance is not limited to specific races of the pathogen), (9) uniform resistance (host resistance is comparable for all races of the pathogen, rather than being good for some races), (10) minor gene resistance (host resistance is controlled by a number of genes with small effects), (11) horizontal resistance (variation in host resistance is primarily due to differences between varieties and between isolates, rather than to specific variety.

- **resistance breeding**
  special crossing and selection methods in order to improve the inherent capacity of a crop plant to prevent or retard the development of an infectious disease resistance.

- **resolving power**
  the ability of an experimental technique to distinguish between two genetic conditions (typically discussed when one condition is rare (e.g. recombination between very closely linked loci) and of particular interest.)

- **respiration**
  oxidative reactions in cellular metabolism involving the sequential degradation of food substances and the use of molecular oxygen as a final hydrogen acceptor.

- **rest**
  a condition of a plant in which growth cannot occur, even though temperatures and other environmental factors are favourable for growth dormancy.

- **rest period**
  rest dormancy.

- **resting bud**
  hibernaculum.

- **resting spore**
  a spore germinating after a resting period (frequently after overwintering), as does an oospore or a teliospore.

- **restitution nucleus**
  a nucleus with an unreduced chromosome number.

- **restorer**
  an inbred line that permits restoration of fertility to the progeny of male sterile lines to which it is crossed.

- **restorer gene**
  a gene and/or allele that is able to restore fertility of a sterile genotype. While genes for sterility frequently belong to the mitochondrial genome (i.e., cytoplasmic), the restorer genes are very often found to belong to the nuclear complement. They are used in hybrid variety production.

- **restorer line (R LINE)**
  a pollen parent line, it contains the restorer gene or genes, which restores cytoplasmic male sterile plants to pollen fertility. It is crossed with an A line in the production of hybrid seeds.

- **restriction analysis**
  determination of the number and size of the DNA fragments produced when a particular DNA molecule is cut with restriction endonucleases.

- **restriction digest**
  the result of the action of a restriction endonuclease on a DNA sample.

- **restriction endonuclease**
  endonuclease that recognises a particular short DNA sequence which they cleave. They help to protect cells from viral infection and are used in work with DNA.

- **restriction enzyme**
  an endonuclease that will recognise a specific target nucleotide sequence in DNA and break the DNA chain at the target. A variety of these enzymes are known and they are extensively used in genetic engineering.

- **restriction enzyme cutting site**
  a specific nucleotide sequence of DNA at which a particular restriction enzyme cuts the DNA. Some sites occur frequently in DNA (e.g., every several hundred base pairs), others much less frequently (rare- cutter, e.g., every 10,000 base pairs.)

- **restriction enzyme, endonuclease**
  a protein that recognises specific, short nucleotide sequences and cuts DNA at those sites. Bacteria contain over 400 such enzymes that recognise and cut over 100 different DNA sequences.

■ **Restriction Fragment Length Polymorphism (RFLP)**
variation in DNA fragment banding patterns of electrophoresed restriction digests of DNA from different individuals of a species. Often due to the presence of a restriction enzyme cleavage site at one place in the genome in one individual and the absence of that specific site in another individual.

■ **restriction map**
a physical map of a piece of DNA showing recognition sites of specific restriction endonucleases separated by lengths marked in numbers of bases.

■ **restriction site**
a certain nucleotide sequence within the double-stranded DNA, it is recognised by a restriction endonuclease. The enzyme cuts the double strand within the recognition sequence. The restriction sites are usually composed of four to six base pairs and are bilaterally symmetric. Both strands are cut either on exactly opposite positions (blunt ends) or alternated ones (sticky ends). The type of cutting depends on the enzyme used.

■ **Restriction-Enzyme-Mediated Integration (REMI)**
a method of transformation that generates tagged mutations.

■ **restrictive condition**
an environmental condition under which a conditional mutant shows the mutant phenotype.

■ **restrictive temperature**
a temperature at which temperature-sensitive mutants display the mutant phenotype.

■ **resynthesis**
the artificial production of autopolyploids or allopolyploids of naturally occurring autopolyploid or allopolyploid plants by utilisation of the presumable parental species (e.g., it was done in wheat and rapeseed.)

■ **retrotransposon**
retrotransposons are a ubiquitous and major component of plant genomes. Those with long terminal DNA repeats (LTRs, Ty1-copia-like family) are widely distributed over the chromosomes of many plant species.

■ **retrotransposon (retroposon)**
a class of genetic elements that includes retroviruses and transposons that have an in-

termediate RNA stage. A transposon that was created by reverse transcription of an RNA molecule.

- **retrovirus**
  an RNA virus that replicates by first being converted into double-stranded DNA by reverse transcriptase.

- **reverse genetics**
  the experimental procedure that begins with a cloned segment of DNA or a protein sequence and uses this knowledge to introduce programmed mutations (through directed mutagenesis) back into the genome in order to investigate gene and protein function.

- **reverse genetics**
  using linkage analysis and polymorphic markers to isolate a disease gene in the absence of a known metabolic defect, then using the DNA sequence of the cloned gene to predict the amino acid sequence of its encoded protein. In general, a technology aiming at isolating mutants of a given sequence, it is also applied for identification of gene function.

- **reverse mutation**
  the production by further mutation of a premutation gene from a mutant gene. It restores the ability of the gene to produce a functional protein. Strictly, reversion is the correction of a mutation (i.e., it occurs at the same site).

- **reverse transcriptase**
  an enzyme, requiring a DNA primer that catalyses the synthesis of a DNA strand from an RNA template. An enzyme that can use RNA as a template to synthesise DNA.

- **reversion**
  the production of a wild-type gene from a mutant gene. The return of a mutant to the wild-type phenotype by way of a second mutational event.

- **revertant**
  an allele that undergoes reverse mutation or a plant bearing such an allele reverse mutation.

- **RFLP**
  Restriction Fragment Length Polymorphism, variations occurring within a species in the length of DNA fragments generated by a species endonuclease.

- **RFLP mapping**
  a technique in which DNA Restriction Fragment Length Polymorphisms (RFLPs) are used as

reference markers for mapping in relation to known genes or other RFLP loci.

- **rheogameon**
  it refers to species composed of segments with marked morphological divergence but gene exchange takes place between them.

- **rhizoid**
  it refers to hair-like filamentous anchorage or absorbing organ.

- **rhizome**
  a horizontally creeping underground stem that bears roots and leaves and usually persists from season to season.

- **rhizosphere**
  the soil near a living root.

- **rho**
  a protein factor required to recognise certain transcription termination signals in *Escherichia coli*.

- **rhodanese**
  the enzyme is defined biochemically by its ability to transfer sulphurs from thiosulphate to cyanide, yielding thiocyanate. It is found in plants, animals and bacteria.

- **rho-dependent terminator**
  a DNA sequence signalling the termination of transcription. Termination requires the presence of the rho protein.

- **rht gene**
  short-straw mutant.

- **rhytidome**
  bark.

- **rhytmicity**
  periodicity.

- **rib**
  a primary or prominent vein of a leaf.

- **Ribonucleic Acid (RNA)**
  a chemical found in the *nucleus* and cytoplasm of cells, it plays an important role in protein synthesis and other chemical activities of the cell. The structure of RNA is similar to that of DNA. There are several classes of RNA molecules, including messenger RNA, transfer RNA, ribosomal RNA and other small RNAs, each serving a different purpose.

- **ribonucleoprotein**
  a protein composed of pre-rRNAs and ribosomal as well as nonribosomal protein components. One of the nonribosomal proteins, the nucleolin, is considered to play a key role in regulation of rDNA transcription, perisomal synthesis, ribosomal assembly and maturation.

- **ribosomal protein**
  one of the ribonucleoprotein particles that are the sites of translation.

- **ribosome**
  a complex organelle (composed of proteins plus rRNA) that catalyses translation of messenger RNA into an amino acid sequence. Ribosomes are made up of two non-identical subunits each consisting of a different rRNA and a different set of proteins.

- **ribozyme**
  catalytic or autocatalytic RNA. RNA with enzymatic activity, for instance, self-splicing RNA molecules in Tetrahymena.

- **ridge tillage**
  a type of soil conserving tillage in which the soil is formed into ridges and the seeds are planted on the tops of the ridges. The soil and the crop residue between the rows remain largely undisturbed. The practice offers opportunities to reduce crop production costs by banding fertilisers and pesticides and reducing the need for field trips.

- **rifamycin(s)**
  a group of antibiotics that inhibit initiation of transcription in bacteria.

- **rind**
  a thick and firm outer coat or covering (e.g., in watermelon, orange, etc. or the bark of a tree.)

- **ring bivalent**
  an association of two chromosomes with terminal chiasmata on both arms.

- **ring chromosome**
  a (sometimes aberrant) chromosome with no ends (e.g., the chromosome of bacteria), an isochromosome may also form a ring in MI of meiosis.

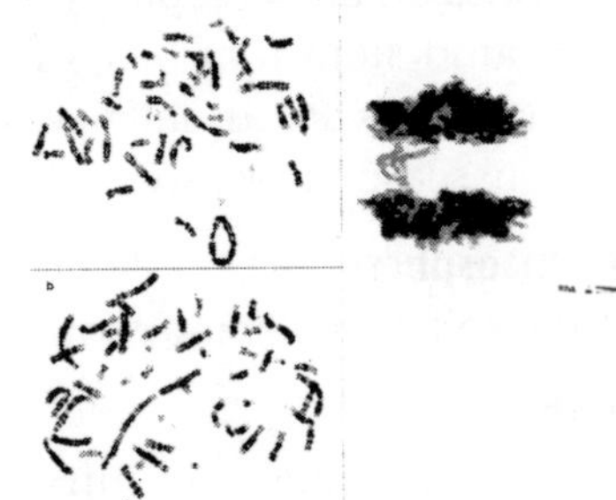

- **ringer solution**
  a physiological saline containing sodium, potassium and calcium chlorides used in physiological experiments for temporarily maintaining cells or organs alive in vitro.

- **ripe**
  mature.

- **RNA editing**
  the insertion of uridines into mRNAs after transcription is

completed, controlled by guide RNA (gRNA). May also sometimes involve insertion of cytidines and possible deletions of bases.

- **RNA in situ hybridisation**
  a technique that is used to identify the spatial pattern of expression of a particular transcript (usually an mRNA). In this technique, the probe is labelled, either radioactively or by chemically attaching a fluorochrome visualised by fluorescence or an enzyme that can convert a substrate to a visible dye. A tissue or organism is soaked in a solution of single-stranded probe under conditions that allow the probe to hybridise to complementary RNA sequences in the cells, unhybridised probe is then removed. Radioactive probe is detected by autoradiography. Fluorochrome is detected by fluorescence microscopy. Enzyme labelled probe is detected by soaking the tissue in the substrate, the dye develops in sites where the transcript of interest was expressed.

- **RNA phage**
  phage whose genetic material is RNA. They are the simplest known phages.

- **RNA polymerase**
  an enzyme that catalyses the synthesis of an RNA strand from a DNA template. The enzyme that polymerises RNA by using DNA as a template. It can also act as a primase initiating DNA replication. (Also known as transcriptase or RNA transcriptase.)

- **RNA replicase**
  a polymerase enzyme that catalyses the self-replication of single-stranded RNA.

- **RNA transcriptase**
  the enzyme responsible for transcribing the information encoded in DNA into RNA. It is also called transcriptase or RNA polymerase.

- **RNAse**
  an enzyme hydrolysing RNA.

- **Robertsonian fusion**
  fusion of two acrocentric chromosomes at the centromere.

- **Robertsonian translocation**
  a chromosomal mutation due to centric fusion or centric fission (i.e., a reciprocal translocation with breakpoints within the centromeric regions) translocation.

- **rogue**
  to remove individuals that have an undesirable phenotype or

that have been shown through progeny tests to have a less desirable genotype from a seed orchard, seed production area or nursery bed; genetic thinning.

■ **rolled paper towelling**
adjusted filter paper or paper towels are used for this method in order to germinate seeds inside and/or between the layers of paper. After germination and growth the viability and/or germability are determined.

■ **roller**
a device that compacts the soil to produce a firm seedbed, like a packer.

■ **rolling-circle replication**
a model of DNA replication that accounts for a circular DNA molecule producing linear daughter double helices.

■ **root**
the lower part of a plant, usually underground, by which the plant is anchored and through which water and minerals enter the plant.

■ **root ball**
the roots and soil or soil mix that they are growing in when lifted from the open ground.

■ **root cap**
a cap of cells covering the apex of the growing point of a root and protecting it as it is forced through soil.

■ **root crop**
a crop, such as beets, turnips or sweet potatoes, grown for its large, edible roots.

■ **root culture**
the in vitro growth of roots (e.g., root tips or root meristem on a synthetic medium).

■ **root cuttings**
root cuttings are made by cutting off pieces of root and planting them under sui conditions. In this way some plants species or varieties can be easily propagated.

■ **root grafting**
the process of grafting scions (shoots) directly on a small part of the root of some appropriate stock. The grafted root then being potted.

■ **root hair**
a tabular outgrowth of an epidermal cell of a root, which functions to absorb water and nutrients from the soil.

■ **root nodule**
a small, gall-like growth on the roots of certain types of plants (legumes), the nodules develop

as a result of infection of the root by bacteria.

■ **root pruning**
cutting the roots of large plants, mainly trees and shrubs, to force more vigorous growth or to prepare the plant for transplantation or transportation.

■ **root sucker**
a shoot arising adventitiously from a root of a plant, mostly at some distance from the main trunk.

■ **rooting**
the natural or induced process of root formation.

■ **rooting compound**
a powdery substance into which fresh cuttings are dipped before inserting in soil or medium, containing hormones, such as kinetins, to encourage root growth.

■ **rootstock**
synonymous with 'rhizome', in horticulture, the bottom or supporting root used to receive a scion in grafting rhizome.

■ **rootstock variety**
in horticulture, there are special (fruit) tree varieties (often of wild-type character) that serve as rootstock for graftings. Usually they show good root formation, resistance traits and compatibility with the scion.

■ **rosarium**
since Roman times, a rose garden and breeding site of roses.

■ **rosette**
an arrangement of leaves radiating from a root crown near the earth.

■ **rot**
to deteriorate, disintegrate, fall or become weak due to decay.

■ **rotary hoe**
an implement that breaks the soil with a circular motion.

■ **rotation**
crop rotation.

■ **rouging**
a manual removal of infected or inferior specimens from an otherwise healthy crop of plants.

■ **row spacing**
the distance between rows of crop plants. It depends on needs for optimal plant growth, plant density, weed control and harvest technology.

■ **rRNA (ribosomal RNA)**
a class of RNA molecules, coded in the nucleolar organiser, that have an integral (but poorly understood) role in ribosome structure and function. RNA

components of the subunits of the ribosomes.

- **rubbing**
smooth the surface of multigerm seed of sugarbeet.

- **rubisco**
a CO_2-fixing enzyme, the key enzyme in photosynthesis, it is the most frequent protein on earth. It has a unique double function of being both a carboxylase and an oxygenase. When acting as an oxygenase, it catalyses the light-dependent uptake of O_2 and the formation of CO_2 in a complicated process (photorespiration), which takes place concomitantly in three organelles—chloroplasts, peroxisomes and mitochondria. There are projects to manipulate the enzyme in order to create an artificial plant that can contribute to reduction of CO_2 content in the atmosphere (i.e., decreasing the so-called greenhouse effect).

- **ruderal plant**
a plant that is associated with human dwellings or agriculture or one that colonises waste ground.

- **rudimentary**
incompletely developed.

- **run out**
separation of nucleic acid or protein molecules by gel electrophoresis.

- **runner**
a procumbent shoot that takes root, forming a new plant that eventually is freed from connection with the parent by decay of the runner, it serves as a vegetative propagule (e.g., in strawberry) stolon.

- **rush**
reed.

- **russet**
a brownish roughened area on the skin of fruits as a result of cork formation.

- **s (svedberg unit)**
a unit of sedimentation velocity, commonly used to describe molecular units of various sizes (because sedimentation velocity is related to size).

- **s0**
a symbol used to designate the original selfed plant.

- **S1 nuclease**
a nuclease that cuts single-stranded DNA and RNA, used for S1 protection experiments in transcript mapping.

■ **S1, S2, S3, etc.**
the representation for continued selfing (self-fertilisation) of plants. S1 designates the generation obtained by selfing the parent plant, S2 the generation obtained by selfing the S1 plant, etc.

■ **saccharose**
a sweet, crystalline substance, $C_{12}H_{22}O_{11}$, obtained from the juice or sap of many plants (e.g., from sugarcane and sugarbeet.)

■ **sacrificial crop**
crop planted to distract pests safely.

■ **sagittate**
arrow-shaped.

■ **salicin**
glucoside.

■ **salicylic acid**
a phenolic substance, which is accumulated in many plants as part of the defence response. It is shown, in some plants, to be necessary for the accumulation of pathogenesis-related proteins. Treatment of plants with salicylic acid can induce resistance.

■ **saline soil**
a soil containing enough salts to reduce plant fertility crop rotation.

■ **s-allele**
an allele of a gene controlling incompatibility in many allogamous plants. Alleles present in both style and pollen are referred to as matching S-alleles. S-alleles usually belong to a series of multiple alleles.

■ **salmon procedure**
a method for producing haploids in hexaploid wheat. 'Salmon' is name of an alloplasmic wheat variety carrying a 1RS.1BL chromosome translocation together with cytoplasm of *Aegilops kotchyi* the interaction of cytoplasmically genetic determinants with, possibly, a parthenogenesis-inducing gene on chromosome arm 1RS of rye results in haploid progeny.

■ **saltation**
a mutation occurring in the asexual state of fungal growth, especially one occurring in vitro culture.

■ **saltational speciation**
quantum speciation.

■ **samara**
a fruit similar to an achene except that the entire seed coat is tightly fused with the pericarp (e.g., ash, elm, tree of heaven, etc.).

- **sample**
  a finite series of observations taken from a population.

- **sample size**
  the number of experimental units on which observations are considered. It may be less than the number of observations in a data-set, due to the possible multiplying effects of multiple variables and/or repeated measures within the experimental design.

- **sampling**
  the method by which a representative sample is taken from a seed lot or something else.

- **sampling distribution**
  the distribution of frequencies with which various possible events could occur or a probability distribution defined by a particular mathematical expression.

- **sampling error**
  variability due to the limited size of the sample.

- **sanger sequence**
  'plus and minus' or 'primed synthesis' method, DNA is synthesised so it is radioactively labelled and the reaction terminates specifically at the position corresponding to a given base.

- **sanitation**
  plant disease control involving removal and burning of infected plant parts and decontamination of tools, equipment, etc.

- **sap**
  the exudate from ruptured tissues emanating from the vascular system or parenchyma.

- **saponin**
  any member of a class of glycosides that form colloidal solutions in water and foam when shaken. It occurs in many different plant species. In cereals, only oats are known to produce these compounds. In oats, the resistance to infection by the take-all fungus, *Gaeumannomyces graminis* var. *tritici*, has been attributed to the family of antifungal saponins known as avenacins, which are present in roots.

- **sapwood**
  the living, softer part of the wood between the inner bark and the heartwood.

- **SAR**
  Scaffold Attachment Region - the position along eukaryotic DNA where it is anchored to the central scaffold of the chromosome.

- **sarment**
  a slender running stem runner.

▪ **sarmentous plant**
runner.

▪ **satellite**
a terminal section of a chromosome, separated from the main body of the chromosome by a narrow constriction.

▪ **satellite DNA**
a highly repetitive DNA, composed of repeated hepta- to deca-nucleotide sequences. DNA of different buoyant density, a minor DNA fraction that has sufficiently different base composition from the bulk of the DNA in order to separate distinctly during cesium chloride density gradient centrifugation, it can derive from nucleus, plastid or mitochondrial DNA.

▪ **saturation mutagenesis**
induction and recovery of large numbers of mutations in one area of a genome or in one biological function, in the hope of identifying all the genes in that area or affecting that function.

▪ **sat-zone**
the secondary constriction of a satellite chromosome nucleolar zone.

▪ **savannah**
a plain characterised by coarse grasses and scattered tree growth.

▪ **scab**
a general term for any unrelated plant disease in which the symptoms include the formation of dry, corky scabs.

▪ **scaffold**
the eukaryotic chromosome structure remaining when DNA and histones have been removed, made from nonhistone proteins. The central framework of a chromosome to which the DNA solenoid is attached as loops, composed largely of topoisomerase.

▪ **scald**
a necrotic condition in which tissue is usually bleached and has the appearance of having been exposed to high temperatures.

▪ **scale**
any thin, scarious body, usually a degenerated leaf.

▪ **scalping**
the removal of material larger than the crop seed during the processing of seeds.

▪ **scanning electron microscopy**
a microscope used to examine the surface structure of biological specimens. A three-dimensional screen image is acquired through focusing secondary electrons emitted from a sample

surface bombarded by an electron beam.

- **scanning hypothesis**
  proposed mechanism by which the eukaryotic ribosome recognises the initiation region of an mRNA after binding the 5' capped end of it. The ribosome scans the mRNA for the initiation codon.

- **scarification**
  the process of mechanically abrading a seed coat to make it more permeable to water. This process may also be accomplished by brief exposure to strong acids (sulphuric acid), it may enhance germination.

- **SCE**
  Sister Chromatid Exchange.

- **schiave**
  collectively refers to grapevine cultivars presently grown on the Southern and Northern slopes of the Eastern Alps and bearing different names, such as Schiava, Trollinger, Rossara, Rossola, Geschlafene, Gansfüsser, Urban, etc. Their common origin has been suggested by historic, linguistic and ampelographic considerations, however, a dendrogram constructed from an AFLP analysis of the 33 Schiave cultivars shows different and in some cases relevant degrees of genomic dissimilarity. The analysed cultivars cluster into at least five taxonomic groups with specific geographic distribution along the valleys of Valtellina, Bergamo and Brescia and those of South Tyrol and Swabia, it is concluded that the common definition 'Schiave' refers to a similar cultivation practice in contiguous regions rather than to a common genetic background.

- **Schiff's reagent**
  a reagent consisting of fuchsin bleached by sulphurous acid that produces a red colour upon reaction with an aldehyde. It is used for chromosome staining.

- **schizocarp**
  a dry, two-seeded fruit of some plants that separates at maturity along a midline into two mericarpes. Each mericarp has a dry, indehiscent pericarp enclosing a loose-fitting ovule (e.g., carrot).

- **scientific method**
  a procedure used by scientists to test hypotheses by making predictions about the outcome of an experiment before the experiment is performed. The results provide support or refutation of the hypothesis.

scion
a portion of a shoot or a bud on one plant that is grafted onto a stock of another.

scion rooting
covering a low graft with soil so that the plant develops roots directly from both the rootstock and the scion.

scission
fission.

sclerenchyma
tissue composed of cells with thickened and hardened walls.

sclerophyllous
having leaves stiffened by sclerenchyma.

sclerotium
a dense, compact mycelial mass capable of remaining dormant for extended periods.

scorch
'burning' of leaf margins as a result of infection or unfavorable environmental conditions.

scorpiod cyme
a determinated inflorescence in which the lateral buds on one side are suppressed during growth, resulting in a curved or coiled arrangement.

screening
examining the properties, performance responses of individuals, lines, genotype or other taxa under an assortment of conditions in order to evaluate the individuals or groups. A routine testing for particular properties.

screening technique
a technique to determine the genotype or phenotype of an organism.

scutellum
a shield-shaped organ of the embryo of grasses, it is often viewed as a highly modified cotyledon in monocots.

scutiform
platter-shaped.

SDS
Sodium Dodecyl Sulphate-polyacrylamide.

SDS gel electrophoresis
(Sodium Dodecyl Sulphate-Polyacrylamide) (SDS-PAGE): in SDS-PAGE, SDS masks protein charge and separations depending only on size as compared to common gel electrophoresis.

secalotricum
a cross combination of rye *(Secale)* and wheat *(Triticum)* in which rye serves as the donor

of the cytoplasm (mother plant), as opposed to triticale.

- **second division**
second meiotic division, which is a mitotic division of chromosomes.

- **second division segregation (SDS)**
the allele arrangement (2+2+2+2) in the spores of Ascomycetes with ordered spores that indicates a crossover between a locus and its centromere. A pattern of ascospore genotypes for a gene pair showing that the two alleles segregate into different nuclei only at the second meiotic division, as a result of a single crossover between that gene pair and its centromere, can only be detected in a ordered ascus.

- **secondary crop**
a crop that originated as a weed of a primary crop (e.g., rye) , in agronomy, a crop grown after a primary crop.

- **secondary gene pool**
species in the secondary gene pool include those from which genes can be transferred to the cultivated species, however with more difficulties as compared to species of a primary gene pool.

- **secondary infection**
any infection caused by inoculum produced as a result of a primary or a subsequent infection and/or an infection caused by secondary inoculum.

- **secondary pairing**
the association of bivalents in polyploids due to genetic, evolutionary or structural factors. By any reason those bivalents appear in groups. Sometimes it seems that the bivalents of a certain genome are closer together than at random or the bivalents of a certain genome occupy certain domains (spatial order) within the meiotic cell (prometaphase and metaphase).

- **secondary root**
lateral root.

- **secondary spermatocyte**
the product of the first meiotic division in male animals.

- **secondary structure**
the alpha-helical or beta-sheet configuration of the polypeptide backbone of a protein.

- **secondary tiller**
it can arise from the prophyll node and leaf node of the primary tillers in cereals. In the same manner, tertiary tillers

may occasionally be produced by secondary tillers. The primary tillers are usually the smallest of the tillers that emerge.

- **second-site mutation**
the second mutation of a double mutation within a gene. In many cases, the second-site mutation suppresses the first mutation, so that the double mutant has the wild-type phenotype.

- **section cutting**
sections are cut from a block of wax around a plant material, usually by microtome.

- **sector**
an area of tissue or colony whose phenotype is detectably different from the surrounding tissue or colony phenotype.

- **sectorial chimera**
a chimera in which the distinct meristem is cross-sectional present, like sectors of a circle chemera.

- **sedimentation**
the sinking of a molecule under the opposing forces of gravitation and buoyancy.

- **seed**
a mature ovule consisting of an embryonic plant together with a store of food, all surrounded by a protective coat.

- **seed bank**
a place or storage in which seeds of rare plants or obsolete varieties are kept, usually vacuum-packed and under cold conditions in order to prolong their viability.

- **seed coat**
the protective covering of a seed usually composed of the inner and outer integuments.

- **seed conditioning**
for marketing, seeds are usually cleaned, sized, treated with fungicides, insecticides or inoculant and finally bagged.

- **seed divider**
a device that divides a seed lot and puts subsamples directly into various number of planting envelopes.

- **seed dormancy**
See **dormancy**.

- **seed drill**
See **drill**.

- **seed flat**
See **flat**.

- **seed health test**
specific tests to determine the absence or presence of certain micro-organisms, known to cause economic loss to crop yield, are carried out in numerous forms. Seeds are generally

surface sterilised and placed onto agar plates or other substrates, which are known to promote the growth of the disease. The percentage of seeds exhibiting the disease are used as a measure of seed health.

- **seed incompatibility**
  a postgamous sterility due to failure of tissue development involved in the formation of the seed.

- **seed increase**
  increase.

- **seed index**
  the 100-g-weight of seeds thousand-grain weight.

- **seed leaf**
  cotyledon.

- **seed lot**
  seeds of a particular crop gathered at one time and likely to have similar germination rates and other characteristics.

- **seed mixture**
  either seed of more than one kind of cultivar or a combination of seed of two or more species.

- **seed multiplication**
  all methods required to grow plants to maturity and produce seeds, including those practices necessary for harvesting, processing and preparing seeds for subsequent plantings.

- **seed orchard**
  plantation of fruit or forest trees, assumed or proven genetically to be superior. It is isolated in order to reduce pollination from genetically inferior outside sources. It is managed to improve the plants and produce frequent, abundant and easily harvested seeds.

- **seed parent**
  the strain from which seed is harvested in the hybrid seed field, also commonly used to designate the female parent in any cross-fertilisation.

- **seed plant**
  an individual plant that is or was used for seed production and/or maintaining the genotype.

- **seed potato**
  a potato tuber that is used for the next growing season in order to produce next generation for selection and experimental testing.

- **seed processing**
  the operations involved in preparing harvested seed for multiplication or marketing.

- **seed production area**
in forestry and horticulture, a stand or plantation designated for collection of seeds for reforestation purposes. It may be rogued of inferior trees and treated in such a manner as to produce large quantities of seed. The wood harvest is usually also an important consideration and the establishment and management is similar to commercial stands. Usually seed production was not an initial consideration at establishment.

- **seed quality control**
control of physiological, sanitary and genetic seed quality characteristics.

- **seed regulation**
the total set of rules and protocols related to variety development and release, seed production, quality control and delivery.

- **seed set**
the process of producing seeds after flowering.

- **seed source**
the location where a seed lot was collected, usually defined on an eco-geographic basis by distance, elevation, precipitation, latitude, etc.

- **seed spacing**
population density.

- **seed stack**
the erect stem on a plant that produces flowers and seed, it is particularly applied to root crops and leafy vegetable crops that produce seed after the desired product (root, head, leaves) has fully developed.

- **seed stand**
any stand used as a source of seed.

- **seed stock**
seed used as a source of germplasm for maintaining and increasing seed of crop varieties; stock seeds.

- **seed trap**
a device for catching the seeds falling on a small area of ground, from trees or shrubs. It is set for determining the amount of seedfall and the time, period, rate and distance of dissemination.

- **seed vessel**
the pericarp (wall of the ripened ovary), which contains the seeds.

- **seed viability testing**
all methods to determine the potential for rapid uniform emergence and development of normal seedlings under both

favourable and stress conditions.

- **seed vigour**
seed properties that determine the potential for rapid uniform emergence and development of normal seedlings under both favourable and stress conditions.

- **seedbed**
a plot of ground prepared for seeds or seedlings.

- **seed-borne pathogens**
carried on or in seeds, for example, in wheat, the streak mosaic virus of barley, the fungi, such as snow mould *(Fusarium nivale)*, *Septoria* spike blotch *(Septoria nodorum)*, *Helminthosporum* leaf blotch or spot blotch *(Cochiobolus sativus, Helminthosporum sativum, syn Bipolaris sorokiana, Drechslera sorokiana)*, loose smut (*Ustilago nuda* or *U. tritici*), common smut or stinking smut *(Tiletia caries)*, dwarf smut *(Tiletia controversa)*, in barley, the stripe mosaic virus, the fungi, such as snow mold *(Fusarium nivale)*, leaf stripe disease *(Pyrenophora graminea)*, *Helminthosporum* leaf blotch *(Helminthosporum gramineum, syn Drechslera graminea)*, net blotch disease *(Pyrenophora teres, syn Helminthosporum teres, syn Drechslera teres)*, loose smut *(Ustilago nuda)*, black smut *(Ustilago nigra)*, hard smut *(Ustilago hordei)*, in rye, the streak mosaic virus of barley, the fungi, such as snow mould *(Fusarium nivale, syn Griphosphaeria nivalis)*, *Septoria* spike blotch *(Septoria nodorum)*, stalk bunt *(Urocystis occulata syn Tuburcinia occulata)*, ergot *(Claviceps purpurea)*, in oats, the streak mosaic virus of barley, the fungi, such as loose smut *(Ustilago avenea) and* in maize, the fungi, such as common smut *(Ustilago maydis)*, seed rots (*Fusarium* spp., *Penicillium* spp.), seedling rots (*Pythium* spp., *Fusarium* spp., *Helminthosporum* spp., *Penicillium* spp., *Rhizopus* spp., *Rhizoctonia* spp., *Deploida* spp.).

- **seeding lath**
commonly a wooden device for obtaining uniformly spaced drills in a seedbed and aiding the even distribution of hand-sown seed in them.

- **seeding machine**
drill.

- **seedling**
a young plant grown from seed.

■ **seedling guard**
a row cover to protect seeds indoors or out.

■ **seed-tree method**
a method of regenerating a forest stand in which all trees are removed from the area except for a small number of seed-bearing trees that are left singly or in small groups.

■ **segmental allopolyploids**
a partial homology or so-called homoeology of chromosome sets combined in an allopolyploid.

■ **segmentation**
the process by which the correct number of segments are established in a developing segmented animal.

■ **segregation**
the separation of alleles during meiosis so that each gamete contains only one member of each pair of alleles.

■ **segregation distortion**
the distortion of the 1:1 segregation ratio produced by a heterozygote it can arise because of abnormalities of meiosis, which results in an *Aa* individual producing an unequal number of *A* and *a* bearing gametes or it may arise from *A* and *a* bearing gametes being unequally effective in producing zygotes.

■ **segregation, rule of**
Mendel's first principle describing how genes are passed from one generation to the next.

■ **segregational load**
genetic load caused when a population is segregating less fit homozygotes because of heterozygote advantage.

■ **segregational petite**
a petite that in a cross with wild-type produces 50% petite and 50% wild-type progeny. Caused by a nuclear mutation.

■ **selec marker**
a physiological or morphological character, which may easily be determined as marker for its own selection or for selection of other traits closely linked to that marker.

■ **selection**
the process determining the relative share allotted to individuals of different genotypes in the propagation of a population. Natural selection occurs if zygotic genotypes differ with regard to fitness.

■ **selection after flowering**
a selection that is only possible after flowering since the critical selective characters are expressed after flowering time, seed and/or fruit formation pe-

riod (e.g., grain size, spike length, fruit colour, etc.).

- **selection coefficient**
a measure of the disadvantage of a given genotype in a population.

- **selection criteria**
the specific characters and plant reactions on which the selection is focussed during the breeding cycles.

- **selection differential**
the difference between the mean of a population and the mean of the individuals selected to be parents of the next generation.

- **selection gain**
in artificial selection, the difference in mean phenotypic value between the progeny of the selected parents and the parental generation.

- **selection intensity**
the ratio of the number of genotypes selected divided by the number of genotypes tested.

- **selection limit**
the exhaustion of genetic variance in a population, so that no further selection response can be expected.

- **selection pressure**
the effectiveness of natural selection in altering the genetic composition of a population over a series of generations.

- **selection prior to flowering**
a selection that is possible before flowering since the critical selective characters are already expressed (e.g., seedling resistance, tillering capacity, head size in cabbage, etc.).

- **selection progress**
the difference between the mean of a population and the mean of the offspring in the next generation born to selected parents.

- **selection response**
the difference between the mean of the individuals selected to be parents and the mean of their offspring. It is expressed by the formula $R = h^2 \times S$ (h^2 = heritability, S = selection coefficient = phenotypic difference between the mean of all selected fractions and the mean of total population.)

- **selection-mutation equilibrium**
an equilibrium allele frequency resulting from the balance between selection against the allele and mutation recreating this allele.

- **selective advantage**
an advantage for survival of a genotype in a population and

for production of viable progeny as compared to other genotypes, which may show a selective disadvantage with respect to fitness and viability.

- **selective agent**
an environmental or chemical agent that imposes a lethal or sublethal stress on growing plants or portions thereof in culture, enabling selection of resistant or tolerant individuals.

- **selective culture medium**
See **selective agent.**

- **selective disadvantage**
inferior fitness of one genotype compared to others in the population.

- **selective fertilisation**
the nonrandom participation of male or female gametes or different genotypes in the formation of zygotes and/or hybrids.

- **selective gametocide**
a treatment that inactivates certain gametes, such as one that produces male sterility but does not affect the female gametes.

- **selective herbicide**
a herbicide that acts against either monocots or dicots, against weeds and not against crop plants or even against species weeds.

- **selective medium**
a culture medium that is enriched with a particular substance to allow the growth of particular strains of organisms.

- **selective neutrality**
a situation in which different alleles of a certain gene confer equal fitness.

- **selective system**
an experimental technique that enhances the recovery of specific (usually rare) genotypes.

- **self**
the fusion of male and female gametes from the same individual.

- **self-assembly**
the ability of certain multimeric biological structures to assemble from their component parts through random movements of the molecules and formation of weak chemical bonds between surfaces with complementary shapes.

- **self-fertilisation**
fertilisation in which the two gametes are from the same individual.

- **selfish DNA**
a segment of the genome with no apparent function other than to ensure its own replication.

- **semiconservative replication**
the mode by which DNA replicates. Each strand acts as a template for a new double helix. The established model of DNA replication in which each double-stranded molecule is composed of one parental strand and one newly polymerised strand.

- **semisterility (half sterility)**
the phenotype of individuals heterozygotic for certain types of chromosome aberration, expressed as a reduced number of viable gametes and hence reduced fertility. Nonviability of a proportion of gametes or zygotes.

- **sense strand**
see **coding strand**.

- **Sequence Tagged Site (STS)**
any site in a chromosome or genome that is identified by a known unique DNA sequence. STSs can be used to form genetic maps by standard mapping procedures.

- **sequencing**
determination of the order of *nucleotides (base sequences)* in a DNA or *RNA* molecule or the order of *amino acids* in a *protein*.

- **SER**
serine (an amino acid).

- **sex chromosome**
a chromosome whose presence or absence is correlated with the sex of the bearer, a chromosome that plays a role in sex determination. Heteromorphic (different shaped, e.g., X and Y) chromosomes whose distribution in a zygote determines the sex of the organism.

- **sex linkage**
the location of a gene on a sex chromosome.

- **sex switch**
a gene, normally found on the Y chromosome in mammals, that directs the indeterminate gonads toward development as testes.

- **sex-controlled trait**
traits that appear more of time in one sex than in the other.

- **sexduction**
sexual transmission of donor *Escherichia coli* chromosomal genes on the fertility factor. A process whereby a bacterium gains access o and incorporates foreign DNA brought in by a modified F factor during conjugation.

- **sex-lethal**
a gene in Drosophila, located on the X chromosome, that is a

sex switch directing development toward femaleness when it is in the on state. It is regulated by numerator and denominator elements that act to influence the genic balance equation (X/A).

- **sex-limited trait**

trait expressed in only one sex. It may be controlled by sex linked or autosomal loci.

- **sex-ratio phenotype**

a trait in Drosophila whereby females produce mostly if not only daughters.

- **sexual selection**

the forces determined by mate choice acting to cause one genotype to mate more frequently than another genotype.

- **Shine-Delgarno hypothesis**

a proposal that prokaryotic mRNA is aligned at the ribosome by complementarity between the mRNA upstream from the initiation codon and the 3' end of the 16S rRNA.

- **shotgun technique**

cloning a large population of different DNA fragments, known to contain a fragment of interest, as a prelude to selecting or screening for that one particular clone containing the fragment of interest for intensive study.

- **shuttle vector**

a vector (e.g., a plasmid) constructed in such a way that it can replicate in at least two different host species (e.g., a prokaryote and a eukaryote). A DNA recombined into such a vector can be tested or manipulated in several cell types.

- **sigma factor**

the protein that gives promoter-recognition specificity to the RNA polymerase core enzyme of bacteria.

- **signal hypothesis**

the major mechanism whereby proteins that insert into or cross a membrane are synthesised by a membrane-bound ribosome. The first thirteen to thirty-six amino acids synthesised, termed a signal peptide, are recognised by a signal recognition particle that draws the ribosome to the membrane surface by interaction with a docking protein. The signal peptide may later be removed from the protein.

- **signal sequence**

the N-terminal sequence of a secreted protein, which is required for transport through the cell membrane.

- **signal transduction cascade**
refers to a series of sequential events, such as protein phosphorylations, consequent upon binding of ligand by a transmembrane receptor, that transfer a signal through a series of intermediate molecules until final regulatory molecules, such as transcription factors, are modified in response to the signal.

- **silent mutation**
mutation in which the function of the protein product of the gene is unaltered.

- **sine**
short interspersed element. A type of small dispersed repetitive DNA sequence (e.g., Alu family in the human genome) found throughout a eukaryotic genome.

- **single- gene disorder**
hereditary disorder caused by a mutant allele of a single (e.g., Duchenne muscular dystrophy, retinoblastoma, sickle cell disease.)

- **single-strand binding protein**
protein that binds to single-stranded DNA usually near the replication fork to stabilise the single strands.

- **Sister-Chromatid Exchange (SCE)**
an event similar to crossing over that can occur between sister chromatids at mitosis or at meiosis.

- **site-specific recombination**
a crossover event, such as the integration of phage lambda, that requires homology of only a very short region and uses an enzyme specific for that recombination. Recombination occurring between two specific sequences that need not be homologous, mediated by a specific recombination system.

- **solenoid structure**
the supercoiled arrangement of DNA in eukaryotic nuclear chromosomes produced by coiling the continuous string of nucleosomes (about 7 nucleosomes per turn.)

- **somatic cell**
a cell that is not destined to become a gamete, a cell whose genes cannot be passed on to future generations.

- **somatic doubling**
a disruption of the mitotic process that produces a cell with twice the normal chromosome number.

- **somatic hypermutation**
  the occurrence of a high level of mutation in the variable regions of immunoglobulin genes.

- **somatic mutation**
  a mutation occurring in any cell that is not destined to become a germ cell. If the mutant cell continues to divide, the individual will come to contain a patch of tissue of genotype different from the cells of the rest of the body.

- **somatic-cell genetics**
  asexual genetics, involving study of somatic mutation, mitotic crossing-over and segregation.

- **somatic-mutation theory**
  a theory to account for the high degree of antibody variability. It suggests that mutation of a basic immunoglobulin gene accounts for all the different types of immunoglobulins produced by B lymphocytes.

- **SOS box**
  the region in the promoter of various genes that is recognised by the LexA repressor. Release of repression results in the induction of the SOS response.

- **SOS repair**
  the error-prone process whereby gross structural DNA damage is circumvented by allowing replication to proceed past the damage through imprecise polymerisation.

- **SOS response**
  repair systems (RecA, uvr) induced by the presence of single-stranded DNA that usually occurs from postreplicative gaps caused by various types of DNA damage. The RecA protein, stimulated by single-stranded DNA, is involved in the inactivation of the LexA repressor thereby inducing the response.

- **southern blotting**
  transfer by absorption of DNA fragments separated in electrophoretic gels to membrane filters for detection of specific base sequences by radiolabelled complementary probes.

- **spacer DNA**
  regions of non-transcribed DNA between transcribed repeated genes such as ribosomal RNA genes in eukaryotes. Its function is probably to do with ensuring the high rates of transcription associated with these genes.

- **specialised (restricted) transduction**
  the situation in which a particular phage will transduce only specific regions of the bacterial chromosome. Form of trans-

duction based on faulty looping out by a temperate phage. Only loci neighbouring the attachment site can be transduced.

- **speciation**
  a process whereby over time one species evolves into a different species (anagenesis) or whereby one species diverges to become two or more species (cladogenesis.)

- **species**
  a group of organisms belong to the same biological species if they are capable of interbreeding to produce fertile offspring. However the biological test of a species is not always available and so there is also a morphological species concept based on anatomical similarities.

- **specific-locus test**
  a system for detecting recessive mutations in diploids. Normal individuals treated with mutagen are mated to testers that are homozygous for the recessive alleles at a number of specific loci, the progeny are then screened for recessive phenotypes.

- **splicing**
  the reaction that removes introns and joins together exons in eukaryotic nuclear primary RNA transcripts.

- **spontaneous mutation**
  a mutation occurring in the absence of mutagens, usually due to errors in the normal functioning of cellular enzymes.

- **sporophyte**
  the diploid sexual-spore-producing generation in the life cycle of plants. The stage in which meiosis occurs. The stage of a plant life cycle that produces spores by meiosis and alternates with the gametophyte stage.

- **stabilising selection**
  a type of selection that removes individuals from both ends of a phenotypic distribution thus maintaining the same distribution mean.

- **stacking**
  the packing of the flattish nitrogen base-pairs at the centre of the DNA double helix.

- **staggered cuts**
  the cleavage of two opposite strands of duplex DNA at points near one another.

- **standard deviation**
  the square root of the variance.

- **standard error of the mean**
  the standard deviation divided by the square root of the sample

size. It is the standard deviation of a sample of means.

- **stasipatric speciation**
  instantaneous speciation caused by polyploidy.

- **statistic**
  a computed quantity characteristic of a population, such as the mean.

- **statistical distribution**
  the array of frequencies of different quantitative or qualitative classes in a population.

- **statistics**
  measurements of attributes of a sample from a population.

- **stem-loop structure**
  a lollipop-shaped structure formed when a single-stranded nucleic acid molecule loops back on itself to form a complementary double helix (stem) topped by a loop.

- **steroid receptor**
  a family of related proteins that act as transcription factors when bound to their cognate hormone ligands. Not all members of this family actually bind to steroids, the name derives from the first family member that was discovered, which was a steroid hormone receptor.

- **stochastic**
  a process with an indeterminate or random element as opposed to a deterministic process that has no random element.

- **strain**
  a pure-breeding lineage, usually of haploid organisms, bacteria or viruses.

- **stringent factor**
  a protein catalysing the formation of an unusual nucleotide (guanosine tetraphosphate, ppGpp) during the stringent response under amino acid starvation conditions.

- **stringent response**
  a translational control mechanism of prokaryotes that represses tRNA and rRNA synthesis during amino acid starvation.

- **structural gene**
  a gene encoding the amino acid sequence of a protein. Non-regulatory gene.

- **submetacentric chromosome**
  a chromosome whose centromere lies between its middle and its end but closer to the middle.

- **subvital gene**
  a gene that causes the death of some proportion (but not all) of the individuals that express it.

- **sum rule**
  the probability that one or the other of two mutually exclusive events will occur is the sum of their individual probabilities. The rule that states that the probability of the occurrence of mutually exclusive events is the sum of the probabilities of the individual events.

- **supercoil**
  a closed double-stranded DNA molecule that is twisted on itself.

- **supergene**
  several loci, which usually control related aspects of the phenotype, in close physical association.

- **supersuppressor**
  a mutation that can suppress a variety of other mutations, typically a nonsense suppressor.

- **suppressive petite**
  a petite that in a cross with wild-type produces progeny of which variable non-Mendelian proportions are petite.

- **suppressor gene**
  a gene that, when mutated, apparently restores the wild-type phenotype to a mutation at another locus.

- **suppressor mutation**
  a mutation that counteracts the effects of another mutation. A suppressor maps at a different site than the mutation it counteracts, either within the same gene or at a more distant locus. Different suppressors act in different ways.

- **swot**
  strengths, weaknesses, opportunities and threats (analysis).

- **sympatric speciation**
  speciation in which the evolution of reproductive isolating mechanisms occurs within the range and habitat of the parent species. This form of speciation may be common in parasites.

- **synapsis**
  the point-by-point pairing of homologous chromosomes during zygotene or in certain Dipteran tissues (e.g., Drosophila salivary glands) that undergo endomitosis. Close pairing of homologues at meiosis.

- **synaptonemal complex**
  a proteinaceous complex that apparently mediates synapsis during zygotene stage and then disintegrates. A complex structure that unites homologous chromosomes during the prophase of meiosis.

- **syncytial blastoderm**
  in insects, the syncytial stage of blastoderm preceding the formation of cell membranes around the individual nuclei of the early embryo.

- **syncytium**
  a single cell with many nuclei.

- **synteny**
  all loci on one chromosome are said to be syntenic (literally on the same ribbon). Loci may appear to be unlinked by conventional genetic tests for linkage but still be syntenic.

- **synteny test**
  a test that determines whether two loci belong to the same linkage group (i.e., are syntenic) by observing concordance (occurrence of markers together) in hybrid cell lines.

- **synthetic medium**
  a chemically defined substrate upon which organisms are grown.

- **T chromosome**
  a chromosome in which a terminal (T) region shows neocentric activity.

- **T1 (generation)**
  a term used in plant genetics, it refers to the progeny resulting from self-pollination of the primary transformant regenerated from tissue culture.

- **T4 phage**
  a type of a bacteriophage used as a source of commonly used ligase, DNA polymerase and polynucleotide kinase.

- **tabular root**
  the main, downward-growing root of a plant, which grows deeply and produces lateral roots along its length.

- **tail**
  single-strand DNA extension added by terminal deoxynucleotidyl transferase.

- **tailings**
  partly threshed material that has passed through the coarse shakers or straw walkers and is eliminated at the rear of a threshing machine.

- **take-all**
  saponin take-all disease.

- **take-all disease**
  a fungal disease *(Gaeumannomyces graminis, syn Ophiobolus graminis)* that attacks wheat plant roots. It causes dry rot and premature death of the plant. Certain strains of *Brassica* plants and *Pseudomonas* bacteria act as natural antifungal agents against the fungus.

- **tandem duplication**
adjacent identical chromosome segments.

- **tandem repeat**
a chromosomal mutation in which two identical chromosome segments lie adjacent to each other, with the same gene order. The DNA, which codes for the rRNA, contains many tandem repeats.

- **tandem repeat sequences**
multiple copies of the same base sequence on a chromosome, used as a marker in ical mapping.

- **tandem selection**
in the case of successive multiple selection the selection concerns other traits in the first few generations than in later generations.

- **tannin**
a generic term for complex, nonnitrogenous compounds containing phenols, glycosides or hydroxy acids, which occur widely in plants (e.g., in the testa of cocoa and beans).

- **tapesia yallundae**
sexual stage of *Pseudocercosporella herpotrichoides* white leaf spot eyespot disease.

- **tapetal cell**
See **tapetum.**

- **tapetal layer**
See **tapetum.**

- **tapetum**
a layer of cells, rich in food, which surrounds the spore mother cells.

- **tapping**
driving spouts into the trunks of maple or pine trees to let the sap out for food and industrial use.

- **taproot**
a large, descending, central root tabular root.

- **taq (DNA) polymerase**
a DNA-dependent RNA polymerase from phage T7, which recognises a very specific promoter sequence. It is used in many expression vectors.

- **taraxum type**
diplospory where the spore mother cell enters the meiotic prophase but because of synapsis there is no pairing and the univalents remain scattered over the whole spindle. The first meiosis results in a restitution nucleus, the second meiosis results in an unreduced dyad.

- **target population**
the target population is the entire group of individuals a breeder is interested in, the group about which he or she

wishes to draw conclusions by several means.

- **target theory**
a theory that predicts response curves based on the number of events required to cause the phenomenon. Used to determine whether point mutations are single events.

- **targeted gene knockout**
the introduction of a null mutation in a gene by a designed alteration in a cloned DNA sequence that is then introduced into the genome by homologous recombination and replacement of the normal allele.

- **tassel**
something resembling this, as at the top of a stalk of maize.

- **tasseling (of maize)**
the time or process when maize emerges tassels tassel detasseling.

- **tata box**
a canonic DNA sequence, part of a plant promoter, promotes the transcription of DNA.

- **tautomeric shift**
the spontaneous isomerisation of a nitrogen base to an alternative hydrogen-bonding form, possibly resulting in a mutation. Reversible shifts of proton position in a molecule. Bases in nucleic acids shift between keto and enol forms or between amino and imino forms.

- **taxis**
a change of direction of locomotion in a motile cell, made in response to certain types of external stimulus, such as temperature, light, nutrients etc.

- **taxon**
a group of organisms of any taxonomic rank (i.e., family, genus, species, etc.).

- **t-DNA**
a portion of the Ti plasmid that is inserted into the genome of the host plant cell.

- **t-DNA**
ti plasmid.

- **te**
tris-EDTA buffer.

- **technology transfer**
the process of converting scientific findings from research laboratories into useful products by the commercial sector.

- **teliospore**
a thick-walled resting spore produced by rust and smut fungi.

- **telium**
pustule containing teliospores.

■ **telocentric (chromosome)**
the chromosomal centromere lies at the end.

■ **telochromomere**
a chromomere that is terminally located chromomere.

■ **telomerase**
an enzyme that adds telomeric sequences to the ends of eukaryotic chromosomes. No template is necessary.

■ **telomere**
the ends of linear chromosomes that are required for replication and stability. The tip (or end) of a chromosome.

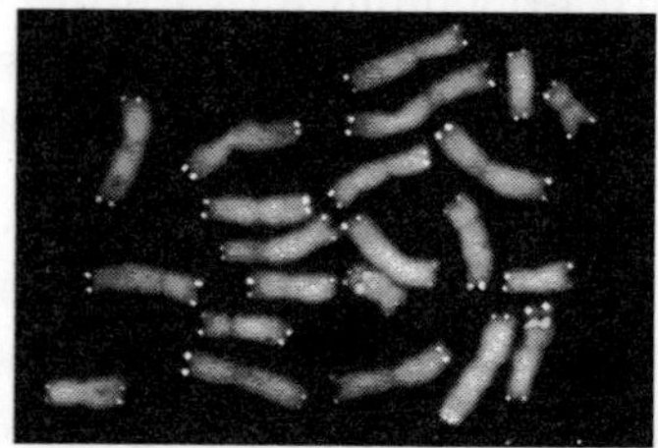

■ **telophase**
the terminal stage of mitosis or meiosis in which chromosomes uncoil, the spindle breaks down and cytokinesis usually occurs. The late stage of nuclear division when daughter nuclei re-form.

■ **telosome**
shorthand of telochromosome, a chromosome with a terminally located centromere.

■ **telotrisomic**
in allopolyploids, such as hexaploid wheat, a cell or individual with one missing chromosome but having a telocentric and an isochromosome for the same arm of the missing one.

■ **temperate phage**
a phage that can enter into lysogeny with its host. A phage that can become a prophage.

■ **temperature-sensitive mutant**
an organism that has a wild-type phenotype at a permissive temperature but a mutant phenotype at a restrictive (non-permissive) temperature.

■ **temperature-sensitive mutation**
a conditional mutation that produces the mutant phenotype in one (restrictive or non-permissive) temperature range and the wild-type phenotype in another (permissive) temperature range.

■ **template**
the DNA single strand, complementary to a nascent RNA or DNA strand, that serves to specify the nucleotide sequence of the nascent strand.

■ **t-end**
a chromosome showing a terminal centromere.

■ **tendril**
part of a stem, leaf or petiole that is modified as a delicate, commonly twisted, thread-like appendage, it is an aid to climbing (e.g., in pea).

■ **tepal**
one of the perianth members in those flowers where there is no distinction between calyx and corolla.

■ **teratogen**
an agent that interferes with normal development.

■ **terminal association**
in meiosis, the achiasmatic association of homologous chromosomes just by unspecific end-to-end attachments.

■ **terminal centromere**
telosome.

■ **terminal redundancy**
a linear DNA molecule with the same sequence (genetic information) at each end.
If genetic information is represented by
ABCDEFGH
then a terminally redundant sequence can be for instance
ABCDEFGHAB.
Terminal redundancy is seen in some phages (e.g., T2) and is generated because a phage head is capable of containing a DNA molecule larger than the complete genome and packaging of DNA into phage heads is determined by the headful. These phages also show circular permutation.

■ **terminalisation**
a progressive shift of chiasmata from their sites of origin to more terminal positions.

■ **terminase**
an ensyme of phage lamda, which generates the staggered cuts at the 'cos' sites during packaging lamda phage.

■ **termination**
the incorporation of the final amino acid into a polypeptide chain and the release of the complete chain from the ribosomes during protein biosynthesis.

■ **termination codon**
stop codon.

■ **terminator**
a nucleotide sequence that acts as a signal for the termination of transcription.

■ **terminator seeds**
a descriptive term used for seeds that have been genetically engineered to produce a crop whose first generation produces sterile seeds, thus preventing a second generation

from being grown from seeds saved from the first. It might be a way to build patent protection directly into high-value, genetically engineered crop variety and thus recoup high research investment costs.

- **terminator sequence**
  a sequence in DNA that signals termination of transcription to RNA polymerase.

- **terpene**
  a hydrocarbon that is composed of two or more isoprene units. They may be linear or cyclic molecules or combinations of both and include important biological compounds, such as vitamins A, E and K growth inhibitor.

- **tertiary gene pool**
  gene transfer from a species of the tertiary gene pool to the cultivated species, since the primary gene poòl usually requires special crossing and embryo rescue techniques in order to get viable hybrids.

- **tertiary structure**
  the further folding of a protein bringing alpha-helices and beta-sheets into three-dimensional arrangements. The folding or coiling of the secondary structure to form a three-dimensional molecule.

- **tertiary tiller**
  secondary tiller.

- **tertiary trisome**
  a chromosome present in addition to the normal diploid complement but as a result of a reciprocal interchange between two standard chromosomes, frequently it occurs in the progeny of a translocation heterozygote balanced tertiary trisomic.

- **test cross**
  a cross between a heterozygote of unknown genotype and an individual homozygous for the recessive genes in question. In general, each cross that contributes to the solution of an experimental question by using more or less defined crossing partners.

- **test mating**
  test cross.

- **test of significance**
  significance test.

- **testa**
  the seed coat, it derives from secondary outgrowths of the nucellus and later the ovule. They are called collars or integuments, the inner and outer integuments become the testa of the mature ovule. It is commonly composed of cuticle,

palisade layer, sandwich cells and parenchym cells caryopsis seed coat.

- **test-cross**
  the crossing of an organism, with an unknown genotype, to a homozygous recessive organism (tester). A cross between an individual of unknown genotype or a heterozygote (or a multiple heterozygote) to a homozygous recessive individual.

- **tester**
  an individual homozygous for one or more recessive alleles and used in a test-cross.

- **tetraallelic**
  in tetraploids, when multiple alleles loci all have different (e.g., four alleles, *A1A2A3A4*).

- **tetrad**
  four homologous chromatids in a bundle during the first meiotic prophase and metaphase. In meiosis, the four haploid cells resulting from a single diploid cell during gametogenesis.

- **tetrad analysis**
  the use of tetrads (definition 2) to study the segregation of chromosomes and genes during meiosis.

- **tetraparental mouse**
  a mouse that develops from an embryo created by the experimental fusion of two separate embryos (blastulas) from matings of different parents.

- **tetraploid**
  having four sets of chromosomes in the nucleus.

- **tetraploidisation**
  the mitotic or meiotic procedure in order produce tetraploids.

- **tetrasome**
  a chromosome present four times tetrasomic.

- **tetrasomic**
  having one or more chromosomes of a complement represented four times in each nucleus.

- **tetrasomy**
  the state of having one or more chromosomes as four copies.

- **tetratype (t)**
  a tetrad containing four different genotypes, two parental and two recombinant. A spore arrangement in Ascomycetes that consists of two parental and two recombinant spores indicating a single crossover between two linked loci.

- **tetrazolium test**
  a quick test to determine seed viability, tetrazolium is a class of chemicals that have the ability to accept hydrogen atoms from

dehydrogenase enzymes during the respiration process in viable seeds. It is the basis of the test during which the tetrazolium chemical undergoes a colour change, usually from colorless to red (formazan). The method was developed in Germany in the early 1940s by G. Lakon. The test is used throughout the world as a highly regarded method of estimating seed viability. It can be completed in only a few hours.

- **TGW**
thousand-grain weight.

- **thc**
hemp.

- **theca**
usually referring to the pollen sac in flowering plants or the capsule in bryophytes.

- **theobromine**
a mutagenically active purine analogue.

- **thermoperiodism**
in some plants (e.g., *Chrysanthemum,* tomato), the floral induction is accomplished by repeated exposure to low night temperatures, separated by periods of higher temperature.

- **thermophil(ic) (plants)**
plants preferring moderate temperature and/or those that cannot cope with low temperatures or frosts.

- **thermotaxis**
taxis.

- **theta structure**
an intermediate structure formed during the replication of a circular DNA molecule.

- **thin**
in a cultivated crop, to remove some plants in order to increase the area available to others.

- **thiol**
any of a class of odiferous, sulphur-containing compounds.

- **thorn**
a hard, sharp outgrowth on a plant (e.g., a sharp-pointed aborted branch.)

- **thousand-grain weight (TGW)**
equals 1,000-grain weight given in grams, it refers to a measure for seed weight and thus indirectly for seed size. From a seed lot 1,000 seeds are randomly taken and get balanced in grams.

- **THR**
threonine (an amino acid).

- **three-parent cross**
three-way cross.

- **three-point cross**
a series of crosses designed to determine the order of three,

nonallelic, linked genes upon a single chromosome on the basis of their crossing-over behaviour three-point test cross.

- **three-way hybrid**
  a hybrid between an inbred line and single cross hybrid.

- **threonine (THR)**
  an aliphatic, polar alpha-amino acid.

- **threshing**
  breaking the seeds free from the seedpods and other fibrous material or the separation of seed from chaff.

- **threshing machine**
  a device that breaks the seeds free from the seedpods and other fibrous material or the separation of seed from chaff.

- **threshold value**
  a critical value on an underlying scale of liability above which individuals manifest a trait or disease.

- **thrips**
  any of several minute insects of the order *Thysanoptera* that have long, narrow wings fringed with hairs and that infest and feed on a wide variety of weeds and crop plants.

- **thylakoid**
  one of the membranaceous discs or sacs that form the principal subunit of a granum in chloroplasts.

- **TI**
  Transcript Imaging.

- **tigella**
  a short stem.

- **tillage**
  the preparation of soil for seeding, it includes manuring, ploughing, harrowing and rolling land or whatever is done to bring it to a proper state.

- **tiller(s)**
  shoots, some of which will eventually bear spikes, which arise from the base of the stem in the grasses.

- **tillering**
  tiller.

- **tillering node**
  a node on the base of the stem in grasses from which shoots arise.

- **tilth**
  the condition of the soil after preparation for seeding.

- **time components**
  basic operations take time, e.g., recombination, time before the test, testing time and time after the test. Breaking of cycling time into components can be used in optimisation of breeding strategies, e.g., choice of the

most efficient testing or selection method.

- **time isolation**

an isolation practice to prevent crossing, it is used when distance isolation is a problem,. caging is too costly or troublesome or when only growing a couple of varieties in a season. It works with any two varieties or species that shed pollen over a limited time and have sufficiently different rates of maturation. To use time isolation, planting dates of two similar varieties are staggered so that by the time the later of the two varieties is flowering, the earlier variety has already finished flowering and is no longer producing or receptive to pollen. The earlier, faster-maturing crop is planted a couple or three weeks before the later, slower-maturing one.

- **tissue**

a group of cells with similar origin and structurally organised into a functional unit. The organs of multicellular organisms are made up of combinations of tissues, of one or more types of cells.

- **tissue culture**

the maintenance or growth of tissue in vitro in a way that may allow further differentiation, preservation or regeneration.

- **titer**

the amount of a standard reagent necessary to produce a certain result in a titration.

- **tolerance**

the ability of a plant to endure attack by a pathogen without severe loss of yield.

- **tolerant**

the ability of a host plant to develop and reproduce fairly efficiently while sustaining disease.

- **tonoplast**

a membrane that borders the vacuole of a cell.

- **topcross**

a cross between a selection, line, clone, etc. and a common pollen parent, which may be a variety, inbred line, single cross, etc. The common pollen parent is called the top cross or tester parent. In maize, a top cross is commonly called an inbred-variety cross. Usually, it is used in order to test the 'general combining ability'

- **topcross progeny**

progeny from outcrossed seed of selections, clones or lines to a common pollen parent.

- **topcross test**
  See **topcross.**
- **top-dressing**
  fertilisation when crop plants are already developed, usually before flowering (i.e., an additional fertilisation to the basic dressing before sowing or close behind). In horticulture, using material, such as compost or manure, that is applied to the surface around the plant to aid in drainage, decrease erosion, prevent moisture loss and keep weeds down.
- **topoisomer**
  topological form of DNA with the same sequence as another but differing in linking number (supercoiling.)
- **topoisomerase**
  enzymes of two types that can remove (or create) supercoiling in duplex DNA by creating transitory breaks in one (type I) or both (type II) strands of the sugar-phosphate backbone.
- **topsoil**
  the fertile, upper part of the soil.
- **torpedo stage**
  the stage of somatic and zygotic embryogenesis in which the embryo or mass of cells are torpedo-shaped.
- **torus**
  the receptacle of a flower.
- **totipotency**
  the ability of a cell to proceed through all the stages of development and thus produce a normal adult.
- **totipotent**
  the state of a cell that can give rise to any and all adult cell types as compared with a differentiated cell whose fate is determined.
- **tough rachis**
  a nonbrittle rachis of a spike. For example, in wheat, the nonbrittleness is determined by two recessive genes on the short arms of chromosomes 3A and 3B. The consequence of threshing is spikelets, rather than grains.
- **toxicity**
  the quality, relative degree or specific degree of being toxic or poisonous.
- **toxigenic**
  See **toxicity.**
- **trabant**
  See **satellite.**
- **trace element**
  an element required only in minute amounts by an organism for its normal growth micronutrient.

- **trachea**
  wood vessel.

- **training**
  the operation of forming (young) tree plants to a wall or espalier or causing them to grow in a desired shape.

- **trait**
  a recognisable quality or attribute resulting from the interaction of a gene or group of genes with the environment character.

- **trans**
  meaning across and referring usually to the geometric configuration of two mutant alleles (ab) across from each other on a pair of homologous chromosomes (Ab/aB).

- **trans conformation**
  in a heterozygote involving two mutant sites (ab) within a gene or gene cluster, the arrangement, Ab/aB.

- **trans-acting**
  referring to mutations of, for example a repressor gene, that act through a diffusable protein product and can therefore act at a distance not simply on the DNA molecule in which they occur.

- **transaminase (aminotransferase)**
  an enzyme that catalyses a transamination reaction.

- **transamination**
  the transfer of an amino group from an amino acid to a keto acid in a reaction catalysed by a transaminase.

- **transcript**
  the RNA product of a gene.

- **transcript imaging (ti)**
  a technique based on AFLP analysis applied to cDNA. It provides a quantified view of all the transcripts in a sample on an electronic image where approximately 20,000 AFLP bands, each representing a transcript, can be detected and quantitated. Differences between strains, samples or treatment can be detected and the differential bands can be sequenced. The sequence of the band leads to the identification of the transcript.

- **transcription**
  the synthesis of RNA using a DNA template. The process whereby RNA is synthesised from a DNA template.

- **transcriptome**
  DNA sequences of the expressed genome.

- **transduce**
  See **transduction**.

- **transduction**
  the movement of genes from a bacterial donor to a bacterial

recipient using a phage as the vector. A process whereby a cell can gain access to and incorporate foreign DNA brought in by a viral particle.

- **transfection**
  the process by which exogenous DNA in solution is introduced into cells. The introduction of foreign DNA into eukaryotic or prokaryotic cells.

- **transfer RNA (tRNA)**
  a class of RNA having structures with triplet nucleotide sequences that are complementary to the triplet nucleotide coding sequences of *mRNA*. The role of tRNAs in protein synthesis is to bond with amino acids and transfer them to the ribosomes, where proteins are assembled according to the genetic code carried by mRNA.

- **transferase**
  an enzyme that catalyses the transfer of a functional group from one substance to another.

- **transformant**
  the cell or individual that was transformed during a transformation procedure.

- **transformation**
  a process by which the genetic material carried by an individual cell is altered by incorporation of exogenous DNA into its *genome*.

- **transformation efficiency**
  the number of bacterial and/or plant cells that uptake and express plasmid DNA and/or a foreign gene, respectively, divided by the mass of plasmid used.

- **transformer**
  an allele in fruit flies (Drosophila) that converts chromosomal females into sterile males.

- **transgene**
  a gene introduced into a host genome by transfection or other similar means.

- **transgenic**
  organisms that have had foreign DNA stably integrated into their genome.

- **transgenic genotype**
  transgenic plants.

- **transgenic organism**
  an organism whose genome has been modified by introduction of novel DNA.

- **transgenic plants**
  a plant that contains an alien or modified DNA (gene) introduced by biotechnological means and which is more or less inherited.

- **transgression**
  segregants in a segregating population that fall outside the variation limit of parental lines.

- **transgressiv(e)**
  transgenic plants.

- **transgressive segregation**
  the segregation of individuals in the F2 or a later generation of a cross that shows a more extreme development of a character than either parent.

- **transient diploid**
  the stage of the life cycle of predominantly haploid fungi (and algae) during which meiosis occurs.

- **transient expression**
  the temporary expression of a gene or genes shortly after the transformation of a host cell (i.e., cells in which the transgene has not been physically incorporated into the genome [s transformation], but is carried as an episome that can be lost.)

- **transition**
  a type of nucleotide-pair substitution involving the replacement of a purine with another purine or of a pyrimidine with another pyrimidine for example GC with AT.

- **transition**
  a type of mutation that involves the replacement in DNA or RNA of one purine with another or of one pyrimidine with another.

- **transition mutation**
  a mutation in which a purine/pyrimidine base pair is replaced with a base pair in the same purine/pyrimidine relationship e.g. GC with AT.

- **translation**
  the process of protein synthesis whereby the primary structure of the protein is determined by the nucleotide sequence in mRNA. The ribosome-mediated production of a polypeptide whose amino acid sequence is derived from the codon sequence of an mRNA molecule.

- **translocase (ef-g)**
  elongation factor in prokaryotes necessary for proper translocation at the ribosome during the translation process. Replaced by eEF2 in eukaryotes.

- **translocation**
  a change in the arrangement of genetic material, altering the location of a chromosome segment. The most common forms of translocation are reciprocal, involving the exchange of chromosome segments between two

nonhomologous chromosomes.

- **transmembrane receptor**
a protein that spans the plasma membrane of a cell, with the extracellular domain of the protein having the ability to bind to a ligand and the intracellular domain having an activity (such as a protein kinase) that can be altered (either increased or decreased) upon ligand binding.

- **transmission**
the spread of a disease agent among individual hosts.

- **transmission genetics**
the study of the mechanisms involved in the passage of genes from one generation to the next.

- **transpiration**
the loss of water vapour from a plant to the outside atmosphere, it takes place mainly through the stomata of leaves and the lenticels of stems.

- **transplant**
to relocate or remove to a new growing place. In biotechnology, the cultured tissue or explant, relocated or transferred to a new site in vitro.

- **transplanting board**
a simple device having regularly spaced slots for the individual plants in order to ensure proper spacing and lining out in the new bed.

- **transposable element**
a chromosomal locus that may be transposed from one spot to another within and among the chromosomes of the genome. It happens through breakage on either side of these loci and their subsequent insertion into a new position either on the same or a different chromosome.

- **transposable genetic element**
a general term for any genetic unit that can insert into a chromosome, exit and relocate, includes insertion sequences, transposons, some bacteriophages and controlling element. A region of the genome, flanked by inverted repeats, a copy of which can be inserted at another place, also called a transposon or a jumping gene.

- **transposition**
in molecular biology, the process of moving a transposon or other inserts from one position to another within a genome translocation.

- **transposon**
a mobile piece of DNA that is flanked by terminal repeat se-

quences and typically bears genes coding for transposition functions.

- **transposon tagging**
  the blocking activity of functional genes by insertion of foreign DNA.

- **transversion**
  a type of nucleotide-pair substitution involving the replacement of a purine with a pyrimidine or vice versa for example GC with TA.

- **transversion mutation**
  a mutation in which a purine pyrimidine replaces a pyrimidine/purine base pair or vice versa e.g. GC with TA.

- **trap nursery**
  sets of plant genotypes are assembled that carry specific resistance to the pathogen in question. It is grown in different geographic locations. It provides information on pathogen populations and may provide resistant genotypes for local breeding.

- **trap plants**
  trap nursery.

- **tree**
  a woody plant, which may grow >10 m tall.

- **tree limit**
  the altitude above sea level at which timber ceases to grow.

- **tree line**
  tree limit.

- **tree ring**
  annual ring.

- **trench planting**
  setting out young trees in a shallow trench or a continuous slit.

- **triazine**
  any of a group of three compounds containing three nitrogen and three carbon atoms arranged in a six-membered ring and having the formula $C_3H_3N_3$. Some of these compounds are used as herbicides.

- **tribe**
  a rank between family and genus, comprising genera whose shared features serve to distinguish them from other genera within the family.

- **trichome**
  a hairy outgrowth on a plant's surface, as a prickle.

- **trigeneric hybrid**
  a spontaneous or experimental hybrid consisting of three genomes of different genera.

- **trihybrid**
  an organism heterozygous at three loci.

- **tri-isosomic**
  in allopolyploids, such as hexaploid wheat, when a cell or individual lacks one chromosome pair while three homologous isosomes for the same arm are present.

- **triplet**
  a unit of three successive bases in DNA and RNA, which code for a specific amino acid.

- **triplet code**
  a code in which a given amino acid is specified by a set of three nucleotides.

- **triplex (type)**
  autotetraploid nulliplex type.

- **triploid**
  a cell having three chromosome sets or an organism composed of such cells.

- **triploidy**
  a state in which three chromosome sets are present.

- **tripping mechanism**
  a pollen dispersal mechanism of some legumes, in which the staminal column is sprung free of the keel and exposed. It can also be initiated by hand in order to imitate insect activity and to stimulate self-pollination, for example, in broad bean *(Vicia faba.)*

- **triradial (chromosome configuration)**
  a chromosomal pairing configuration of three in which the homologous chromosomes are arranged like star.

- **trisome**
  See **trisomic.**

- **trisomic**
  a diploid cell with an extra chromosome. Basically a diploid with an extra chromosome of one type, producing a chromosome number of the form 2n + 1.

- **trisomic analysis**
  a method for mapping gene loci on individual chromosomes by comparing disomic and trisomic segregation patterns of a series of individuals.

- **trisomic series**
  a complete set of trisomics of a given plant species in which all different chromosomes of the complement are available as trisomics in appropriate individuals (e.g., in barley, rye, tomato, maize, etc.)

- **tristyly**
  three lengths of the style relative to the anthers (short, medium, long.)

- **tritium**
  a radioactive isotope of hydrogen.

- **trivalent**
an association of three homologous chromosomes in meiosis.

- **tRNA (transfer RNA)**
transfer RNA. Small RNA molecules that carry amino acids to the ribosome for polymerisation into a polypeptide. During translation the amino acid is inserted into the growing polypeptide chain when the anticodon of the tRNA pairs with a codon on the mRNA being translated.

- **trophic**
tropism.

- **tropism**
a directional response by a plant to a stimulus.

- **TRP**
tryptophan (an amino acid).

- **true breeding**
a situation in which a group of identical individuals always produce offspring of the same phenotype when intercrossed.

- **truncated**
appearing as though abruptly cut across toward the apex.

- **truncation selection**
a breeding technique in which individuals in whom quantitative expression of a phenotype is above or below a certain value (the truncation point) are selected as parents for the next generation.

- **trunk**
stem.

- **truthfully labelled seed**
seed with label of the producer with information on the seed quality.

- **tryphine**
entomophilous.

- **trypsin**
an enzyme of the pancreatic juice, capable of converting proteins into peptone.

- **tryptophan(e) (TRP)**
a heterocyclic, nonpolar, alpha-amino acid.

- **tube nucleus**
vegetative nucleus.

- **tube planting**
setting out young plants in narrow, open-ended cylinders of various materials.

- **tuber**
a swollen stem or root that functions as an underground storage organ.

- **tuber blight**
late blight.

- **tuberous**
bulbous.

- **tuberous roots**
they look like tubers, but are actually swollen, nutrient-storing root tissue (e.g., in dahlias). During the growing season, they put out fibrous roots to take up moisture and nutrients. New growth buds or eyes, form at the base of the stem. This area is called the crown.

- **tubiform floret**
a small flower in a flower head or other cluster showing a tube-like shape.

- **tubular floret**
tubiform floret.

- **tuft**
many stems in a close cluster at ground level, not spreading (e.g., as in some grasses.)

- **tumour suppressor gene**
a gene encoding a protein that suppresses tumour formation. A gene that normally prevents unlimited cell division. When both copies of the gene are lost or mutated the cell is transformed to a cancer phenotype. Examples are the P53, retinoblastoma and Wilm's tumour.

- **tumour virus**
a virus that is capable of inducing a cancer.

- **tunic**
a loose, outer covering or skin surrounding some corms and bulbs (e.g., in onion and tulip.)

- **turf**
the surface of grassy land, consisting of soil or mould filled with the roots of grass and other plants.

- **turgescent**
turgid.

- **turgid**
the crisp, fresh condition found when the cells of the plant are amply supplied with water to the extent that they are fully extended, as opposed to wilted.

- **turgor**
the rigidity of a plant and its cells and organs resulting from hydrostatic pressure exerted on the cell walls.

- **turner syndrome**
an abnormal human female phenotype produced by the presence of only one X chromosome (XO).

- **twig**
a shoot or small branch of a tree or shrub.

- **twin**
a pair of individuals produced at one birth.

■ **twin seedling**
a common feature of plants. Frequently, one of the seedlings is diploid, while the other is haploid via apomictic development (e.g., in asparagus, rye, etc.). In the past, twin seedlings were used for haploid selection.

■ **twin spot**
a pair of mutant sectors within wild-type tissue, produced by a mitotic crossover in an individual of appropriate heterozygous genotype.

■ **two-point cross**
a cross involving two loci.

■ **two-rowed**
barley.

■ **two-strand double crossover**
a double crossover that involves only two of the four chromatids of a tetrad.

■ **type I error**
in statistics the rejection of a true hypothesis.

■ **type II error**
in statistics the accepting of a false hypothesis.

■ **typological thinking**
the concept that organisms of a species conform to a specific norm. In this view variation is considered abnormal.

■ **tyrosinase**
an enzyme that converts tyrosine to dopa and oxidises this to dopa quinone.

■ **tyrosine (TYR)**
an aromatic, polar alpha-amino acid.

■ **U**
Uracil or uridine.

■ **uag stop codon**
amber codon.

■ **ultracentrifugation**
centrifugation carried out at high rotor speeds (<100,000 rpm) and therefore under high centrifugal forces (<750,000 g).

■ **ultracentrifuge**
a set-up for high-speed centrifugation between 65,000-100,000 rounds per minute.

■ **umbel**
an inflorescence in which all the pedicels arise at the apex of an axis.

■ **umbelliferous plants**
tap-rooted plants with minute flowers aggregated into flat or umbrella-shaped heads (e.g., carrot, parsnip, celery, dill, parsley, etc.)

■ **unavailable nutrients**
plant nutrients that are present in the soil but cannot be taken

up by the plant roots because they have not been released from the rock by weathering or from organic matter.

- **unavailable water**
water that is present in the soil but cannot be taken up by the roots because it is strongly adsorbed on to the surface of particles.

- **unbalanced diallelic**
a genotype involving a multiple allelic locus in autotetraploids where two alleles are represented an unequal number of times.

- **unbalanced experimental design**
an experiment or set of data in which all treatments or treatment combinations are not equally represented, a common cause of unbalanced experiments is unequal mortality among entries in a test. See **randomised-block design**.

- **unbalanced translocation**
a type of chromosome translocation in which a loss of chromosomal segments results in a deleterious genetic effect.

- **uncomplete block design**
an experimental design that is preferable where large numbers of cultivars are compared in a single yield trial. The entries in each replication are subdivided into smaller blocks in a manner designed to reduce the error caused by soil variation. Usually, it refers to a lattice design, considering the restriction that (1) the number of testers must be harvested, (2) inferior strains cannot be discarded prior to harvest to reduce harvest expenses and (3) researchers still must analyse the experiment as a lattice design.

- **unconscious selection**
indirect selection by breeders (sometimes called 'parallel selection') when in addition to a target characteristic another trait is taken to the next generation. It can be a morphological, biochemical or physiological trait genetically linked to the target characteristic.

- **underdominance**
a condition in which the phenotypic expression of the heterozygote is less than that of either homozygote.

- **underplant crop**
in horticulture, adding one or more complementary, low-growing plants beneath and around taller plants. In agriculture, main crops can be underplanted with vegetables.

It helps these plants to have some shade in the heat of the summer and one can often get a crop longer than otherwise. Wide ranges of combinations between main and underplant crops are possible and are applied for several purposes.

- **under-replication**
certain heterochromatic chromosome regions and ribosomal DNA that show a slower replication as compared to the remaining genomic DNA.

- **understock**
the bottom or supporting part of a graft composed of either root or stem tissue or both rootstock graft.

- **unequal crossing over**
non-reciprocal crossing over caused by mismatching of homologous chromosomes. Usually occurs in regions of repetitive DNA.

- **unifactorial**
monogenic.

- **uniflorous**
showing one flower only.

- **uniform crossover**
in genetic algorithms, a breeding technique in which it is randomly decided for each element of a breeding pair of individuals whether they should be switched.

- **uniformity**
describes the state of a population or group in which all the individuals are genetically identical. It is a typical feature of clonal varieties (e.g., potato). In general, lack of diversity within and between plant species is apparent in modern cultivars.

- **unilateral**
the type of panicle (e.g., in oats) where the branches are all turned to one side like a pennant.

- **unilateral inheritance**
inheritance that is associated with linkage in sex chromosomes.

- **uninemic chromosome**
a chromosome consisting of one double helix of DNA.

- **uniparental inheritance**
the transmission of certain phenotypes from one parental type to all the progeny. Such inheritance is generally produced by organelle (mitochondria, chloroplast) genes.

- **unipolar (spindle)**
a cell spindle with only one pole.

- **unique DNA**
a length of DNA with no repetitive DNA sequences.

■ **unisexual**
a flower that possesses either stamens or carpels but not both (i.e., a plant possessing only male or female flowers.)

■ **univalent**
a single chromosome observed during meiosis when bivalents are also present, it has no pairing mate.

■ **univalent shift**
a spontaneous change in monosomy from one chromosome to another. It is caused by partial asynapsis or desynapsis during meiosis.

■ **universe**
population.

■ **unloader**
a device on a combine to transport the grains from a bunker to a transporter.

■ **unmixed codon family**
group of four codons sharing their first two bases and coding for the same amino acid.

■ **unreduced gametes**
gametes not resulting from common meiosis and so showing the number of chromosomes per cell that is characteristic of a sporophyte. They spontaneously arise as a consequence of irregular division in anaphase I of meiosis. They may contribute to spontaneous (meiotic) polyploidisation. In rye, they were used for production of tetraploids via valence crosses.

■ **unspecific resistance**
horizontal resistance.

■ **unstable mutation**
a mutation that has a high frequency of reversion. A mutation caused by the insertion of a controlling element, whose subsequent exit produces a reversion.

■ **unusual bases**
other bases in addition to the normal adenine, cytosine, guanine and uracil. Found primarily in tRNAs and produced by post-transcription modification of one of the normal bases.

■ **upgrading**
the reprocessing of a seed lot to remove low quality seeds or other materials. The remaining seeds are of higher quality than the original.

■ **upper palea**
upper glume pale.

■ **upstream**
a convention on DNA related to the position and direction of transcription by RNA poly-

merase (5'>3'). Downstream (or 3' to) is in the direction of transcription whereas upstream (5' to) is in the direction from which the polymerase has come.

- **urea**
  a compound, $CO(NH_2)_2$, occurring in urine and other body fluids as a product of protein metabolism, an important plant fertiliser.

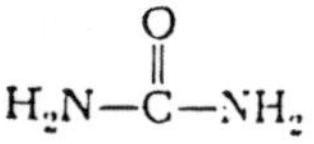

$$H_2N-\overset{\displaystyle O}{\overset{\|}{C}}-NH_2$$

Urea – Mammals

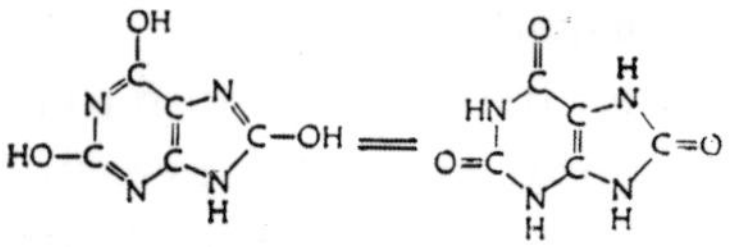

- **uredospore**
  a sexual spore of the rust fungi.

- **v/v**
  may indicate simple proportion (e.g., 3:1 v/v), may indicate percent volume.

- **vacilin**
  globulin.

- **vacuole**
  a transparent vesicle that is usually large and singular in mature cells but small in some meristematic cells. It is filled with a dilute solution that is isotonic with the cytoplasm.

- **valence cross**
  crossing of individuals of different ploidy level. A method, for example, used in production of tetraploid rye varieties by crossing a tetraploid genotype as female and a diploid as male, respectively. Tetraploid F1 seeds could be selected by green-grained xenia on a pale-grained mother plant, caused by fusion of reduced gametes of the mother plant (carrying a recessive allele for pale seed colour) with unreduced gametes of the male parent (carrying the dominant allele for green seed colour).

- **Valine (VAL)**
  an aliphatic, nonpolar amino acid.

- **value-added grains**
  nutrient-enhanced varieties.

- **value-enhanced grains**
  nutrient-enhanced varieties.

- **variability**
  the sum of different genetic or phenotypic characters within different taxa.

- **variable**
  a property that may have different values in various cases.

- **variable region**
  a region in an immunoglobin molecule that shows many se-

quence differences between antibodies of different specificities, the part of the antibody that binds to the antigen.

- **Variable-Number-of-Tandem-Repeats (VNTR) locus**
  locus that is hypervariable because of tandemly repeated DNA sequences. Presumably variability is generated by unequal crossing over or slippage during replication. A chromosomal locus at which a particular repetitive sequence is present in different numbers in different individuals of a population or in the two different chromosome homologues in one diploid individual.

- **variance**
  the average squared deviation about the mean of a set of data. A measure of the variation around the central class of a distribution, the average squared deviation of the observations from their mean value.

- **variant**
  an individual organism that is recognisably different from an arbitrary standard type in that species.

- **variate**
  a specific numerical value of a variable.

- **variation**
  the differences among parents and their offspring or among individuals in a population.

- **variegation**
  the occurrence within a tissue of sectors or clones with differing phenotypes. Patchiness, a type of position effect that results when particular loci are contiguous with heterochromatin.

- **varietal hybrid**
  the product resulting from the mating of two varieties.

- **varietal protection**
  protected variety.

- **variety**
  a plant differing from the other member of the species to which it belongs by the possession of some hereditary traits. Breeding varieties can be classified according to the manner of propagation, such as clone varieties (maintained by vegetative propagation), line varieties (maintained by self-fertilisation), panmictic varieties (propagated by cross-fertilisation) or hybrid varieties (produced by directed crosses).

- **variety listing**
  it refers to a list of new varieties recommended for agricul-

ture and horticulture. After official performance tests, only those varieties that meet the standards are accepted for a national or international variety list. Important for variety listing are the testing for value and use as well as distinctness, homogeneity and stability of the candidates.

- **variety machine**
  a term introduced by N. F. Jensen illustrating his breeding programme. It refers to a segmented time process composed of from 10 to 15 annual input segments of homozygous selections, each of which decreases in number with attrition over time until all of the selections in a segment disappear, either through discard or elevation to parent or variety status. The 'machine' is fed with about 10,000 new selections annually, not much time is needed to spend on the so-called 'annual procedural flow'. Most time should be spent on quality of the material fed into the 'machine', the production of superior fixed lines can be more or less automatic.

- **variety mixture**
  a composite population made up of a random mixture of different varieties. It may exhibit considerable phenotypic variation in one or more characters, but which have one or more desirable agronomic traits in common blend.

- **variety release**
  the official approval of a variety for multiplication and distribution.

- **vascular bundle**
  a discrete, longitudinal strand that consists principally of vascular tissue.

- **vector**
  in DNA cloning, the plasmid or phage chromosome used to carry the cloned DNA segment.

- **vegetation**
  a collection of plants of diverse or the same species.

- **vegetational analysis**
  any of various methods of studying small (sample) areas of constituent plants, often counting the numbers of plants of communities to make extrapolations to a larger area (e.g., estimating the yield.)

- **vegetative**
  applied to a stage or structure that is concerned with feeding and growth rather than with sexual reproduction.

■ **vegetative cone**
in flowering plants, the vegetative cone consists of three different layers. They are called dermatogen (L1), subdermatogen (L2) and corpus (L3). From the dermatogen the epidermis is formed, from the subdermatogen the mesophyll and the gametes and from the corpus the vascular bundle, the flesh, the pith and the adventitious roots. The cell division of the dermatogen and the subdermatogen is usually anticline, in other words, vertical to the layer (in this way the layers show surface expansion), the cell division of the corpus is usually periclinal (i.e., the cells divide parallel to the layers).

■ **vegetative meristem**
gives rise to parts, such as stems, leaves, roots, etc.

■ **vegetative nucleus**
the tube-nucleus of a pollen grain in a flowering plant.

■ **vegetative propagation**
a reproductive process that is asexual and so does not involve a recombination of genetic material (e.g., cloning of potato.)

■ **vegetative stage**
whorl stage.

■ **veinure**
threads of fibrovascular tissue in a leaf or other organ, especially those that branch.

■ **venation**
nervature.

■ **venom**
See **toxicity.**

■ **venomous**
See **toxic.**

■ **ventral**
the inner side, furrowed in the grain of, for example, wheat, barley and in the caryopsis of oats.

■ **ventral furrow**
the groove running along the length of the ventral side of the caryopsis.

■ **vermiculite**
a porous form of mica, a mineral that makes good rooting media for seed germination because of its capacity to retain moisture and permit aeration.

■ **vernalin**
a hypothetical hormone-like substance found in plant meristematic regions. Produced by vernalisation, this substance is apparently graft transmissible, but has not yet been identified. Different cold-requiring species may form different substances during vernalisation.

- **vernalisation**
  the treatment of germinating seeds with low temperatures to induce flowering at a particular preferred time. In the genetic model plant, *Arabidopsis thaliana*, a gene *Flc* suppresses the formation of flowers during cold periods, another gene *Vrn2* triggers that suppression. However, in spring, the suppression is raised in a way that the gene 'remembers' the previous cold period, because the gene *Vrn2* itself is cold-insensitive. There is a similar gene in *Drosophila melanogaster*, which also serves like 'chemical memory'.

- **vernation**
  the arrangement of bud scales or young leaves in a shoot bud.

- **vertical gene transfer**
  outcrossing.

- **vertical resistance**
  the existence of differential levels of resistance to different races of a given pathogen conditioned by one or a few qualitative genes resistance.

- **verticilate**
  whorled.

- **viability**
  the probability that a fertilised egg will survive and develop into an adult organism. The term is also often applied to plant germination experiments with comparisons across phenotypic classes under standard specific environmental conditions.

- **viable**
  capable of germinating, living, growing or sufficiently developed physically as to be capable of living.

- **vibrator separator**
  a machine utilising a vibrating deck for separating seeds on the basis of their shape and differing surface textures.

- **vicilin**
  a protein common in broad bean.

- **video microscopy**
  microscopy that takes advantage of video as an imaging, image-processing or controlling device.

- **vignin**
  globulin.

- **vigour test**
  vigour tests of seed determine the potential ability to develop into normal, healthy plants under a wide range of field conditions. Stress tests, accelerated ageing and other techniques are used to provide a more sensitive index of seed quality. Poor

emergence, dormancy, seed deterioration and other quality factors can be identified through vigour testing.

- **vine**
  climbing plants with woody or herbaceous stems that climb, twist, adhere or scramble over other taller objects. They can climb by tendrils, aerial roots, twining stems, twining leaf-stalks, adhesive disks or hooks.

- **vine-growing**
  viticulture.

- **vine-louse**
  a plant louse *(Phylloxera vitifoliae)* that injures the grapevine.

- **viniculture**
  viticulture.

- **virescence**
  greening of tissue that is normally devoid of chlorophyll (e.g., the abnormal development of flowers in which all organs are green and partly or wholly transformed into structures like small leaves.)

- **virion**
  a virus particle.

- **viroid**
  a piece of infectious nucleic acid. In plant pathology, any of a class of plant pathogenic agents consisting of an infectious, single-stranded, free RNA molecule.

- **virulent phage**
  a phage that cannot become a prophage, infection by such a phage always leads to lysis of the host cell.

- **viruliferous aphids**
  aphids acting as vector for plant viruses.

- **virus**
  a particle consisting of a nucleic acid (RNA or DNA) genome surrounded by a protein coat (capsid) and sometimes also a membrane, which can replicate only after infecting a host cell. A virus particle may exist free of its host cell but is incapable of replicating on its own.

- **virus-free plant**
  a plant that shows no sign of viral particles or symptoms.

- **viscid**
  sticky.

- **vital colouring test**
  vital staining.

- **viticulture**
  the science, culture or cultivation of grapes and grapevines.

- **vitreous grain**
  characterising slightly translucent kernels.

■ **viviparous**
applied to a plant whose seeds germinate within and obtain nourishment from the fruit. It refers also to a plant that reproduces vegetatively from shoots rather than an inflorescence.

■ **vivipary**
viviparity.

■ **vivisection**
the action of cutting into or dissecting a living body.

■ **vivotoxin**
pathotoxin.

■ **v-j joining**
the joining of a variable (V) gene segment and a joining (J) gene segment in the first step of the formation of a functioning immunoglobulin gene.

■ **volunteer**
self-sown cereals volunteer plants.

■ **volunteer plants**
plants that have resulted from natural propagation, as opposed to having been deliberately planted by humans.

■ **vybrid**
the first- and subsequent-generation progenies of crosses of heterozygous facultative apomicts.

■ **W/V**
Weight in Volume, as the number of grams of constituent in 100 ml solution.

■ **Wahlund effect**
a subdivided population contains fewer heterozygotes than predicted despite the fact that all subdivisions are in Hardy-Weinberg equilibrium.

■ **water sprout**
a shoot arising from a bud located on wood that is not older than one year.

■ **water-absorbing capacity**
in breadmaking, a high capacity to absorb water is required. This is associated with hard milling texture, high protein content and the degree of starch damage during the milling process.

■ **waterlogged**
soil saturated with water.

■ **wax coating**
a thin layer covering the stem, leaves, flowers and fruits of most plants. Waxes are manufactured as oily droplets in epidermal cells, from which they migrate to the outer surface of the plant via tiny canalculi in cell walls and crystallise as rods and platelets. Their pattern of deposition is sometimes used as a micromorphological character below the

genus level. The wax coating reduces the water transpiration of the plant and is involved in water balance and resistance mechanisms against diseases.

- **waxiness**
  the phenomenon of whitish, powdery or waxy covering of plant leaves, stems or flowers.

- **Waxy Hull-Less Barley (WHB)**
  a barley mutant that is rich in soluble fibre and low in fat content. These characteristics make it a nutritionally valuable ingredient for food products. It has been shown that the soluble fibre, beta-glucan, reduces cholesterol and lowers blood glucose and insulin response following a meal.

- **waxy maize**
  maize that produces kernels in which the starch that is contained within those kernels is at least 99 percent amylopectin, versus the average of 72-76 percent amylopectin in common starch.

- **weathering**
  all the physical, chemical and biological processes that cause the disintegration of rocks at or near the surface.

- **weed**
  a plant that occurs opportunistically on land that has been disturbed by human activity or under cultivated land where it competes for nutrients, water, sunlight or other resources with cultivated plants.

- **weed killer**
  herbicide.

- **weediness**
  unwanted effects of a plant.

- **weeding**
  remove weeds from the crop stand.

- **western blot**
  western blotting.

- **western blotting**
  a technique for identifying a particular protein using antibodies after electrophoretic separation in a gel and transfer to a membrane.

- **wet milling**
  process in which feed material is steeped in water, with or without sulphur dioxide, to soften the grains in order to help separate the kernel's various components.

- **whorl**
  an arrangement of leaves, etc., in a circle around a stem verticilate.

- **whorl stage**
  the developmental stage of a grass plant prior to the emergence of the inflorescence.

- **wide cross**
  wide hybridisation.

- **wide hybridisation**
  cross combinations between taxonomically remote species or genera.

- **wide hybrids**
  wide hybridisation.

- **wide row planting**
  the method of wide-spaced sowing seeds in multiple rows (e.g., for better selection of young breeding material) or several rows with wide parallel channels for irrigation.

- **wild type**
  the most frequently observed phenotype or the one arbitrarily designated as 'normal'.

- **wilt**
  a type of disease in which wilting is a principal symptom.

- **wilting point**
  the percentage of water remaining in the soil when the plants wilt permanently.

- **wind pollination**
  pollination by wind-borne pollen allogamy.

- **wind-row**
  a loose, continuous row of cut or uprooted plants placed on the surface of the ground for drying to facilitate harvest.

- **wing(s)**
  the two expanded parts of the glume in, for example, wheat, which lie on each side of the keel. In general, a membranous or thin and dry expansion or appendage of a seed or fruit.

- **winged fruit**
  See **wing(s)**.

- **winnower**
  a simple device for seed cleaning from weeds and chaff using an air flow.

- **winnowing mill**
  See **winnower**.

- **winter annuals**
  plants from autumn-sown seed that bloom and fruit in the following spring, then die.

- **winter killing**
  killing frost.

- **winter spore**
  teleutospore resting spore.

- **winter-and-spring wheat**
  facultative growth habit.

winter-type (of growth habit)
plants germinating in autumn, requiring vernalisation during the wintertime for flower induction during the following year.

witch's broom
massed outgrowth of branches of woody plants caused by fungi (e.g., by rusts.)

withertip
death of the leaf beginning at the tip, usually in young leaves.

wobble
the ability of certain bases at the third position of an anticodon in tRNA to form hydrogen bonds in various ways, causing alignment with several possible codons. Referring to the reduced constraint of the third base of an anticodon as compared with the other bases thus allowing additional complementary base pairings.

working collection
a collection of germplasm kept under short-term storage conditions, commonly used by breeders or researchers.

world collection of crop plants
a global collection of samples of a species or genera. It is a coordinated activity of several countries and institutions under the IPGRI.

x
designates the basic number of chromosome sets.

X chromosome
a sex chromosome found in a double dose in the homogametic sex and in a single dose in the heterogametic sex.

X chromosome inactivation
in female mammalian embryos, the early random inactivation of the genes on one of the X chromosomes, leading to mosaicism for functions coded by heterozygous X-linked genes.

X linkage
the inheritance pattern of genes found on the X chromosome but not on the Y.

XA ratio
the ratio between the X chromosome and the number of sets of autosomes.

X-AND-Y linkage
the inheritance pattern of genes found on both the X and Y chromosomes (rare).

xanthophyll
yellowish-brownish (oxygen-containing) carotinoids occurring in the chloroplasts (e.g., the lutein of leaves.)

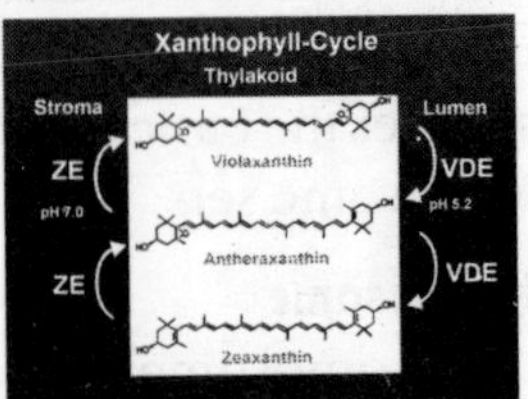

- **xenia**
  a situation in which the genotype of the pollen influences the developing embryo of the maternal tissue (endosperm) of the fruit to produce an observable effect on the seed.

- **xenogamy (cross-pollination)**
  intercrossing between flowers of different individuals, as opposed to geitonogamy.

- **xenotransplant**
  the implantation of an organ or limb from one species to another organism in a different species heterograft.

- **xeric**
  dry.

- **xerograft**
  heterograft.

- **xerophyte**
  a drought-resistant plant or plants that grow in extremely dry areas.

- **xerophytic**
  growing on dry conditions.

- **x-inactivation**
  the repression of one of the two X-chromosomes in the somatic cells of females as a method o dosage compensation. At an early embryonic stage in the normal female, one of the two X-chromosomes undergoes inactivation, apparently at random From this point on all descendant cells will have the same X-chromosome inactivated as the cell from which they arose. Thu a female is a mosaic compose of two types of cells, one which expresses only the paternal X chromosome and another which expresses only the maternal X chromosome.

- **X-Inactivation Centre (XIC)**
  locus on the X chromosome i mammals at which inactivatio is initiated.

- **X-ray crystallography**
  a technique, using X rays, to de termine the atomic structure c molecules that have bee crystallised. A technique fo deducing molecular structur by aiming a beam of X rays at crystal of the test compoun and measuring the scatter c rays.

- **XTA**
  chiasma.

- **XYLAN**
  a polysaccharide of xylos and a component of hemice lulose.

▪ **XYLEM**
a plant tissue consisting of various types of cells that transports water and dissolved substances toward the leaves.

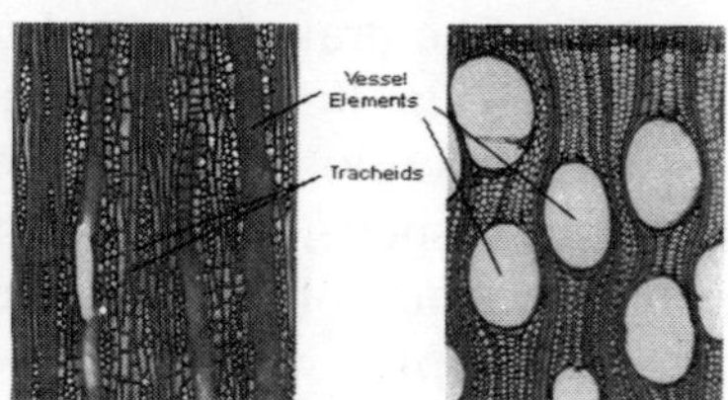

▪ **xylene**
See **xylol.**

▪ **xylol**
a liquid solvent refraction index.

▪ **xylose**
an aldopentose sugar that is commonly found in plants and especially in woody tissue.

▪ **Y linkage**
the inheritance pattern of genes found on the Y chromosome but not on the X (rare).

▪ **YAC**
Yeast Artificial Chromosome, a linear vector into which a large fragment of DNA can be inserted. The development of YAC's in 1987 has increased the number of nucleotides which can be cloned.

▪ **yearling**
a one-year-old seedling and/or plantlet.

▪ **yeast**
a general term for a fungus that can exist in the form of single cells, reproducing by fission or by budding.

▪ **Yeast Artificial Chromosome (YAC)**
originating from a bacterial plasmid, a YAC contains additionally a yeast centromeric region (CEN), a yeast origin of DNA replication (ARS) and two telomere regions (TEL). YACs are capable of cloning very large pieces of DNA.

▪ **yellow sigatoka**
black sigatoka.

▪ **yellows**
a plant disease characterised by yellowing and stunting of the host plant.

▪ **yield**
commonly, the aggregate of products resulting from the growth or cultivation of a crop and usually expressed in quantity per area.

▪ **yield appraisal**
vegetational analysis.

▪ **yield monitoring**
collecting data on the amount of production at regular intervals and by certain means (e.g., GPS readings). The resulting yield map is basic to decisions

about fertilisation, pest control and other adjustments in a system of precision farming.

- **yield structure**
an analysis used to determine the numerous morphological and physiological components of a plant contributing to the final yield (given in different measures).

- **yield trial**
a nursery or experimental design in order to determine the yield capacity of a crop or yield components.

- **Y-junction**
the point of active DNA replication where the double helix opens up so that each strand can serve as a template.

- **Z chromosome**
a sex chromosome that is limited to the male sex.

- **Z DNA**
a left-handed form of DNA found under physiological conditions in short GC segments that are methylated. It may be important in regulating gene expression in eukaryotes.

- **zade method**
long-plot design.

- **zeatin**
a mitogen isolated from maize kernels.

- **zein**
a storage protein of maize found in the endosperm.

- **zeleny test**
a test to measure the protein quality. The grain is milled to form a white flour and mixed with a suspension agent. The resulting suspension volume is then measured in millimetres, for example, wheat with a zeleny volume between 20 and 30 are accep table (< 19 low, 25 medium, 35 high, 45 very high, >50 extremely high).

- **zygote**
the unique diploid cell formec by the fusion of two haploic cells (often an egg and a sperm, that will divide mitotically tc create a differentiated diploic organism.

- **zygotic induction**
when a prophage is passed intc an F- cell during conjugation i may begin vegetative growth The sudden release of a lysogenic phage from an Hf chromosome when the proph age enters the F-(minus) cell and the subsequent lysis of the recipi ent cell.

- **zygotic selection**
the forces acting to cause dif ferential mortality of an organ ism at any stage (other than ga metes) in its life cycle.